MARINE BIOLOGY

Figure 1–1   The coral reefs present an unlimited number of microenvironments for animals to live in.

MARINE BIOLOGY
Second Edition

John Reseck, Jr.
Rancho Santiago College

A RESTON BOOK
Prentice Hall, Englewood Cliffs, N.J. 07632

LIBRARY OF CONGRESS
Library of Congress Cataloging-in-Publication

Reseck, John, 1935–
  Marine biology / John Reseck, Jr.
    p.   cm.
  Bibliography: p.
  Includes index.
  Summary: A basic marine biology textbook, introducing marine
environments and covering the general principles of the field,
including taxonomy and ecology.
  ISBN 0-8359-4454-9
  1. Marine biology. [1. Marine biology.] I. Title.
QH91.R47   1988
574.92—dc19   87-26464

Editorial/production supervision and
    interior design: Margaret Lepera
Cover design: Lundgren Graphics, Ltd.
Cover photo: Courtesy John Reseck, Jr.
Manufacturing buyer: Peter Havens

© 1988, 1980 by Prentice-Hall, Inc.
A Division of Simon & Schuster
Englewood Cliffs, New Jersey 07632

Printed in the United States of America
10 9 8 7 6 5 4 3 2 1

ISBN 0-8359-4454-9

PRENTICE-HALL INTERNATIONAL (UK) LIMITED, *London*
PRENTICE-HALL OF AUSTRALIA PTY. LIMITED, *Sydney*
PRENTICE-HALL CANADA INC., *Toronto*
PRENTICE-HALL HISPANOAMERICANA, S.A., *Mexico*
PRENTICE-HALL OF INDIA PRIVATE LIMITED, *New Delhi*
PRENTICE-HALL OF JAPAN, INC., *Tokyo*
SIMON & SCHUSTER ASIA PTE. LTD., *Singapore*
EDITORA PRENTICE-HALL DO BRASIL, LTDA., *Rio de Janeiro*

To my several thousand students over the last twenty-two years who have formed my thinking on how marine biology should be taught.

Contents

Preface

With the emphasis in public and private schools turning back to basics, the science requirements in many states have been upgraded and increased. In revising this book, the author has made an effort not only to update the marine biological information, but also to add basic biological material to meet the new curriculum standards in general science. These general biological principles are presented by using marine organisms and environments as examples. When logical, these examples are compared with humans and their environment. Studying marine environment provides many chances for problem solving and requires critical thinking and observation by the student. It is an excellent model from which to learn general biology because it not only holds students' interest but also contains all the biological groups of organisms and principles.

John Reseck, Jr.

MARINE BIOLOGY

Figure 1–2 Sponges are a major part of the filter feeding community of the coral reef regions.

Part One

BASIC UNDERSTANDINGS

Figure 1–3 Studying marine biology can be an enjoyable experience. Gathering at tide pools and learning the natural history of their inhabitants is a good way to spend a weekend.

Chapter One

CIVILIZATION AND THE SEA

DEFINITION OF TERMS USED IN CHAPTER 1

Maximum sustainable harvest: The greatest quantity of a population of fish or invertebrates, such as lobster, that can be taken each year and yet totally replace itself through natural reproduction of the species.

Nodule: Usually a small sphirical mass composed of a combination of minerals often containing manganese.

Organic waste: Carbon compounds remaining after raw sewage has been treated for disposal.

Protein deficiency: A dietary deficiency common to the poorer people of the world that causes malfunction and deformation of the body and eventually death.

Treatment plants: Plants where raw sewage is converted to harmless organic substances, which are recycled as fertilizer, etc.

Benthic: Occurring on the bottom of a body of water.

From the beginning of human existence, the sea has had a great influence on the people that lived at its edge. It was probably first used as a source of food and second, for transportation. These two uses are ancient and continuing. What has changed in modern times? What additional uses have humans found for the sea?

RECREATION

The sea as a source of recreation has become not only an important **sociological factor** in the sanity of many city residents, who look forward to weekends and vacation at the beach, but also a major **economic factor.** Property values in coastal cities have increased faster and higher than in other areas. This means more profit for the landowner, more tax money for the government, and, as the population increases, less beach area for the public.

There was a time not many years ago when people complained that the state was buying up most of the beaches. Now the only place many people can get to the beach may be where the state bought it. Private lands are being closed to the public because the public is now too numerous for

Figure 1–4 Spearfishing for large game fish, like this rooster fish, is thrilling, but it can also deplete fish and other populations beyond their natural recovery rate if too few fish are allowed to escape and reproduce.

the landowner to tolerate. Public beaches are crowded and used to capacity. We use the sea to sail, fish, scuba dive, swim, or photograph; or we just lie in the sun while the endless breakers roll to shore. There seems something therapeutic about watching the breakers as they crash on the shore, only to recede and advance again. Like looking into a campfire, watching the waves allows individuals to compare themselves to the forces of nature, in relation to which they are so insignificant, and thereby brings them in contact with themselves. Recreation is a major use of the sea for millions of people around the world and is a major economic consideration for many governments, both state and local.

A PLACE TO DUMP

As human population increased, outhouses filled up, and neighbors were too close to relocate them. Humanity had a problem, a problem that was solved by creating a sewer. The waste was then carried away from where people lived. The question of course, arose as to where to dump it unless the ocean or any large river was near by. Then, it was dumped there. After all, the ocean was so large no one could see across it or even find its bottom. Certainly a little raw sewage could not hurt it. Soon there were more people and more sewage, and eventually someone discovered that raw sewage carried disease. Although treatment plants were developed to stop disease, still the population grew. Even the treated sewage, disease-free if treated properly, is now so voluminous that the organic waste settles out over the floor of the ocean like a gigantic oxygen-eating blanket. As the material decomposes, it consumes oxygen and smothers the **benthic** life of the area. After years of dumping, we find the ocean may give back some of the waste dumped there. Off New York, the dumped waste of many years is starting to move toward shore as a large mass of putrid organic material. Many of the beaches, so necessary for our psychological survival, have been closed, not just in New York, but in California and other states as well. This method of getting rid of our waste needs drastic revision if recreational use of the sea is to continue.

Along with more people came more technology. More technology meant more factories. More factories meant more industrial waste. In general, industrial waste, which is chemical in nature, was different from human waste; it was able to destroy all life in the area where it was dumped. As it contaminated rivers and bays, not only did humans have to give up recreation in them, but also most of the organisms living there were killed. With

Figure 1–5 A pipeline dumps waste water into the ocean. We must take great care not to dump materials that will have long-term effects on the marine life. A break in the food chain caused by a pollutant would destroy the sea as we know it.

chemical pollution we are not just talking about pollution, but about destruction. If society pollutes the seas, it is inconvenient; if it destroys life in the seas, it will cost us life as we now know it on the planet. In recent years this fact has been recognized, and governments are taking steps to forestall this possible disaster. For those who are aware of the problems, the governmental processes seem too slow, even though they are, in reality, very fast for governments. The major concern of the lawmakers is not to make drastic changes so rapidly as to cause the economic collapse of major industries. Because of this economic concern, contaminants such as DDT are outlawed, but users are given one, two, or three years in which to produce a substitute before they must stop use altogether. It is easy to argue logically on either side. Governments have three responsibilities: first, to keep their people from hunger; second, to keep them producing (working); and third, to protect them from disaster (ecological or political, such as war). Sometimes, to the frustration of the environmentalist who can see the large picture, the government generally is more concerned with the smaller picture of food and productivity; after all, success at these two endeavors has historically kept governments in power. The need to maintain an environment in which we can survive is a new concept, one we have never faced until now.

RADIOACTIVE WASTE

One of society's most pressing problems in the late 1980s is the disposal of radioactive waste. The United States has narrowed the selection of waste-disposable sites down to three as of 1987. One site is in Deaf Smith County, Texas; one is in Yucca Mountain, Nevada; and another site is near Richland, Washington. There is an ongoing search to find a site in the deep ocean. It is known that the earth's surface is composed of large plates that move, or drift. The average movement is approximately one inch a year. If we can find the center of one of these large plates there should be very little movement of a kind that could break open a radioactive waste container. We also know the currents in the large ocean basin travel in a circular pattern. In the center of such a circular pattern there is very little water movement to carry away any leakage of radioactive material. It has also been established that the red clay sediments found on the deep ocean floor would act as a good barrier to any leaked radioactive material. What we are searching for is a deep ocean area in the middle of a large geological plate that occurs in the center of a large circular oceanic current, and has a deep layer (several hundred feet) of red clay covering it. A number of such places have been identified. If all research continues to be positive, we will most likely be burying our atomic waste in these areas by the turn of the century. Only future generations will know if it was a wise decision.

TRANSPORTATION

Ever since humans could hang onto a log, the sea has provided a means of transportation. The shipping industry of today carries millions of tons of cargo throughout the world although the day of sea travel for personal transportation has passed. Now that air travel has all but totally replaced it and what took weeks by sea now takes hours by air, the only sea travel in modern times is aboard the cruise ships. This is really less a form of travel than a form of recreation.

The transport of cargo, however, is a huge industry. Well over half of all incoming cargo enters the United States at our seaports. Marine transport

Figure 1–6 Boats require constant maintenance due to the corrosive action of salt water. This maintenance is a major portion of commercial fishing expenses, and pushes the price of seafood even higher.

Figure 1–7 The commercial fishing fleets are important for our food production.

costs only a small fraction of one cent per ton-mile. The larger the ships get, the less the cost per ton. New and larger ships are being built every day to help carry more cargo for less cost, and there is no new technology on the horizon that will change this in the near future. Harbors are being dredged to let the new super tankers and cargo ships enter. Where there is no harbor large enough, pipelines are being laid out to sea so the big oil tankers can load and unload without needing to come into shallow water. If all the shipping around the world were stopped, many people would starve, others would freeze, and millions would be out of a job.

COMMERCIAL FISHING

The most important use of the sea is considered by most to be the food taken from it. The sea yields approximately 12 percent of the animal protein consumed by the world population. Small countries that are surrounded by the sea, such as Japan, Iceland, and the Pacific Islands, take a much larger percentage of their protein from the sea. Countries with large land masses and less coastline per square mile of the total area use a smaller percentage. In the United States, we use seafood to make up about 5 percent of our total animal protein.

In the late 1940s, the annual catch was around 20 million tons of fish and invertebrates. By the 1970s we had more than tripled that amount. Most areas are now being fished, even the cold, rough seas of the Antarctic. The problem we now face is selecting organisms for which to fish. Although the potential of the sea for food has not yet even been scratched, the fishing pressure on a few species has all but destroyed them. The abalone, once abundant on the West Coast of the United States, has been so depleted that the small population remaining will not support a fishery at all. The problem lies in the selection of species for human consumption. People are particular. They like lobster, crab, tuna, cod, and a few other specific types. When we concentrate our efforts on a single species, we soon reduce the population of that species so that it becomes economically unfeasible to continue to fish it. An example is the abalone. Fifteen years ago it was fifty cents a pound; today it is $50.00 a pound. When the catch becomes small enough that the price escalates to a point where the public stops buying it, the industry either goes out of business, or fishes for something else it can catch more of, which it then sells at a more reasonable price. It is this economic rule of survival for those engaged in commercial fishing that stops the total depletion of a particular species.

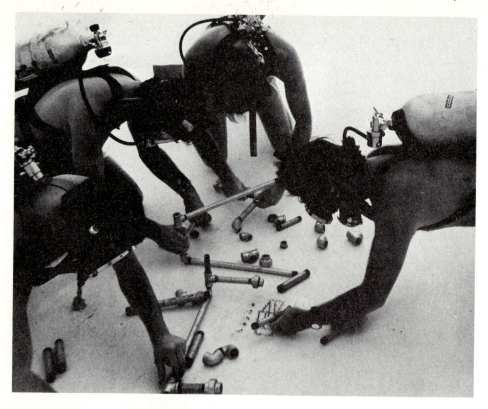

Figure 1–8 Schools are training divers to work underwater. This training in data collecting allows the marine biologist to make first hand observations.

All the latest scientific technology has gone into the creation of fishing fleets. The USSR and Japan, the world leaders in this area, have entire fleets that are self-sufficient for long periods of time at sea. They catch the fish and can it before they return to shore. However, the danger of such a fleet is that it moves into an area and can totally over-fish a population and then just move on. When the area being depleted of fish is off a foreign country's shore, political ramifications can occur. Many countries are setting their boundaries 200 miles out to sea to protect their fish population.

With the help of the scientist, the commercial fishing industry has found out that its fishing must be done scientifically if it is to be continued. With no fishing pressure on a fish population, the number of fish will reach a predictable level of abundance and stay there. The only fluctuation would be due to natural environmental factors, such as availability of food, proper

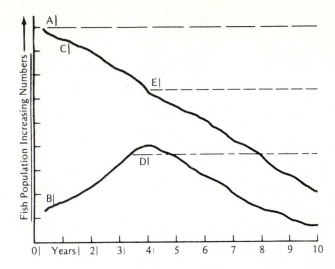

A. Represents normal population of a type of fish with no fishing pressure. Remains relatively stable due to natural limiting factors.
B. Amounts of fish caught by fishing.
C. Population of fish as decreased by the fishing pressure.
D. Represents the maximum sustainable yield that can be continued year after year.
E. Represents the fish population if the maximum sustainable catch is maintained. The fish population will remain stable. The secret is to recognize when the maximum sustainable yield has been reached and not overfish a population.

Figure 1–9 Effects of fishing pressure by commercial fisheries on a given population of fish.

temperature, and the like. If a fishery is developed to take these fish, their population can be maintained if the fishing harvest is small. The mackerel of the North Sea is a good example. If we increase the fishery and take more fish each year, we must be careful not to reduce the population below the ideal point where it can replace all of the fish we take out each year. If we fish at this level, called the *maximum sustainable yield,* we can maintain the greatest possible yield, year after year. If we catch too many, the number of fish will decrease each year until we fish ourselves out of a job. Examples of severely overfished animals are the blue whale of the Antarctic and the halibut of the North Atlantic. Fishing just the correct amount to maintain a maximum annual yield is both a science and an art. Research is constantly

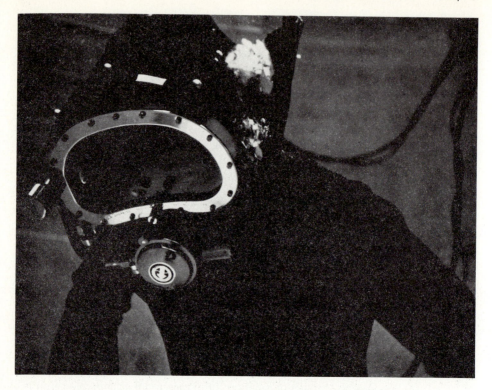

Figure 1–10   The commercial diver plays an important part in the development of various resources such as oil drilling and maintenance of underwater structures.

being done to help us better understand the fish population and how to utilize it to the maximum without depleting the population.

Another problem is that seafood is so expensive that the poor people do not benefit from it. Only the well-to-do can afford to buy it. Because of its cost, seafood does not look like an answer to the world's protein deficiency. Unless we can lower the cost considerably, it will not be available to the poor. One possible answer would be to fish the more common species that are not normally eaten, such as shark. Most shark is good to eat and easy to catch. All we have to do is remove the psychological stigma against eating it. In 1975 few restaurants served shark; in 1988 most do. Many other seafoods could also be utilized. Perhaps when people get hungry enough, these psychological stigmas will be forgotten. During World War II, shark was sold in most markets, even though it was not called shark. It went under several names, such as ocean blue fish, gray fish, or any others the market

Figure 1–11   Tar is quite often found on the rocks or the beaches in many areas of the world. Most of this tar comes from natural oil seepages in the oceans. Oil spills caused by humans increase the tar on the beach for a period of time.

wanted to place on it. It was good eating and people bought it by the ton. When the law changed after the war and the markets had to call it shark, no one bought it. Most of us in the United States are lucky enough to be able to eat with our heads instead of our stomachs. Let us hope that we are always so fortunate as to be able to pick and choose only the choice portions of what is available to eat.

MINING THE SEA

Minerals have been washing into the sea from the great land masses for millions of years. If we could extract minerals from sea water, we would recover nearly any mineral we wanted from it. Sea water consists of dis-

Figure 1–12   The gigantic tankers that carry oil from one port to another are absolutely indispensable to the world economy but they do create a potential hazard to the environment if they should have an accident and create oil spills.

solved molecules from every known element and water-stable compound known to us. As technology develops, we will get more and more of our minerals from the sea. At present we take very little, from gold off Alaska, to iron in Japan, and diamonds offshore from Southwest Africa. With the newer deep exploration methods, we have found nodules on the sea floor containing high concentrations of minerals that could be recovered if the nodules could be brought to the surface economically. We can work on the sea floor now at 500 meters.

As we obtain more resources from the sea, we will become more aware of its importance and the necessity of protecting it. Being the creatures we are, people will pay close attention to the protection of the oceans only when it means money or survival. That time is near. New oil leases are being opened along the upper Alaska continental shelf and in Southern California, as well as along the northeastern shelf of the United States. The need for oil

Figure 1-13 Oil platforms are becoming more common as humans try to keep up with the demand for more energy.

to keep our industry moving has overcome the fear of contaminating the water. We react to financial need before we react to almost any other, with the exception of our actual physical survival. If things continue as they are, we may be faced with that awesome reality within our lifetime.

FRESH WATER

The world is suffering from overpopulation. A close look at this problem reveals at least two major areas of concern: disposing of waste and creating enough food. We have vast unused land where people could live if we could eliminate their waste and feed them. Water is a key factor in the solution of both those problems. There is ample rainfall in the world, but it does not fall in areas where the heavy populations are. We do not build a city where it rains several hundred inches a year if we can help it. People choose, rather, to live in the arid and semiarid parts of the world. By damming rivers and creating long pipelines and aqueducts, we move water from where it is ample to where we live. With increased population, the need for water is beginning to exceed the supply. The Los Angeles water district has had to bring in so much water from the northern Owens Valley that the water table has dropped in the Valley and a once lush area has been turned into an arid stretch of land that looks like a waterless desert. Still, pumps are taking out more water and the Valley will be lost in the near future. The wildlife is leaving or dying; the plants are drying up; and instead of a lush, beautiful valley, only a barren desert is left. This is only one of many examples. We must solve this type of problem.

Fresh water from sea water will be one of the major factors relieving the pressure. Although fresh water can be extracted from the sea in several ways, the cost is higher than we presently want to pay. As the demand grows greater, we will be willing, through necessity, to pay a higher price for our water. When we are, installations will be constructed to augment our fresh water supply from the oceans of the world. Some parts of the world have already done so. As the need increases, we will see new technology and

Figure 1–14 Small coastal freighters deliver supplies to many parts of the world. They are a cheaper method of transportation to remote areas than trucks and in many cases the only way supplies can be brought to small villages.

more efficient ways to accomplish the task. Society is crisis oriented, and given the proper stimulation, that of survival, can accomplish what was previously thought impossible. Fresh water from the sea is already a reality; with a few refinements, it will be a major part of our existence.

ELECTRICITY FROM THE SEA

The earth is in a power crisis. We see the predictable end of much of our fuel in the relatively near future. Oil and coal are both limited, and although we have plenty for half a century, what then? Electrical turbines run by the sea currents and by the tidal flow have been proposed. Both USSR and France have working plants that use the tidal flow principle; however, in most of the world, the tides are not extreme enough to make this a feasible solution for waning fuel supplies. Nuclear power appears to be the only technologically possible answer at this time but it is a highly controversial one, especially after the accidents at Three Mile Island in the USA, and Chernobyl in the USSR. As a source of cooling for the electrical generating plants, the ocean will play a major role. An electric generating plant requires millions of gallons of cool water to circulate through heat exchangers to carry away excess heat and recondense the steam used to turn the turbines. This water is brought in through gigantic pipes from the sea and returned to the sea. There is no chemical pollution, but the returning water, which is considerably warmer than the normal water in the area, causes thermal pollution. This rise in temperature will destroy some native life forms that are not able to tolerate the increased temperature range. In most cases a **succession** will occur in which organisms with a higher temperature tolerance range move in and become part of the community. The area of warm water around the outfall could conceivably be used as a special breeding area for introduced species of commercial value. Such an endeavor would make use of the colder water around the patch of warm water as an invisible fence to keep the commercially valuable species available. Electrical power from nuclear power plants, which rely on the sea for cooling, is already a reality and may very well be of greater worldwide significance in the future.

The meltdown of the power plant at Chernobyl in the USSR has shown the entire world the real dangers that must be guarded against. The Chernobyl meltdown sent a radioactive cloud all the way around the world. A thousand miles away, people could not drink milk from cows that ate contaminated grass. Radioactive rain fell on Canada. For the first time, the whole world paid attention; everyone realized just how small our earth is and how vulnerable we all are.

Figure 1-15 Humans continue to try to adapt to the marine environment.

FARMING OF THE SEA

The underwater vegetable garden of Jules Verne is not likely ever to come about. Limited farming of the sea is a reality and is increasing, although very slowly. At present we farm oysters, abalone, algae, and a few others; but at today's production rates these farms are insignificant for the world food supply.

It has been estimated that we must increase the world food production by 3.5 percent each year to keep up with the increase in population. Society is realizing that the possibilities of expanding agriculture are much more limited than the possibilities of expanding aquaculture. In 1970 aquaculture produced only 2.6 million tons as compared to just ten years later in 1980, when 8 million tons were produced. By 1990 the predictions are that 40 million tons will be produced. At the present time Asia produces about 80 percent of all aquaculture, but this is expected to change as the need and technology spread throughout the world.

Once again the law of supply and demand will dictate the speed at which this resource will be utilized. We know it is possible, but much money and research must be put into making it economical.

REVIEW QUESTIONS

1. How does the recreational use of the sea affect the government?
2. How does biological pollution differ from chemical pollution?

3. What is the "maximum sustainable yield"?

4. What is the importance of the nodules found on the sea floor?

5. Why is fresh water not taken from the sea at a greater rate than is now being extracted?

6. Why does the disposal of radioactive waste pose such a problem for society?

Figure 2–1   The intertidal zone on a rocky shore is a fascinating place to study. When the tide is out, the entire area can be easily observed.

Chapter Two

SUCCESSION

AND A FEW OTHER

BASIC PRINCIPLES

DEFINITION OF TERMS USED IN CHAPTER 2

Algae: A group of plants that in spite of their appearance do not have true roots, stems, or leaves. Generally found in water, such as seaweed.

Anguilla: North Atlantic eel

Balanus: Acorn barnacle

Climax community: A well-established group of organisms that dominate a given area and remain stable inhabitants until some major environmental change takes place.

Clupea: White-bait

Diatoms: One-celled plants

Donax: Bean clam

Dynamic: Used here to mean everchanging.

Eisenia: Palm kelp

Emerita: Mole crab

Evolving: Undergoing a gradual generation-to-generation change in the heredity of an organism.

Genetic: Relating to the hereditary factors that control the makeup of an organism.
Haliotis: Abalone
Impact: Having to do with the effect of external factors on an organism, such as the wave impact (physical force) or the impact of DDT (chemical) on an organism.
Larva: Any early or immature form of an organism.
Ligia: Shore louse
Littorina: Periwinkle
Mytilus: Mussel
Oncorhynchus: Salmon
Phyllospadix: Surf grass
Physiology: The functions (mainly chemical) of a living organism.
Protozoa: One-celled animals
Tivela: Pismo clam
Tolerance: Range within which an organism can endure changes in any of the environmental factors, such as temperature, and survive.

In the study of any biological science, which is, in fact, the study of life, one of the things we must come to grips with is death. In all living communities death is as important as birth because without it there would be no room for those that are born. Each type of living thing needs a special set of circumstances for it to live in a healthy manner. Humans, for example, need food, clothing, and shelter (among numerous other things such as oxygen, water, etc.). Each of these needs has a level or **optimum** at which the individual functions best, as the need of a human for food illustrates. If people get no food, they die. If they get too little food, their physical form changes and they cannot function well. With just the right amount of food, they look good and function well; but with too much food their physical form changes, and again they can't function at their best level of efficiency. This relationship is true with all species and all factors which affect them.

Because the environment (that is, the total of all physical and biological factors which affect an organism) changes, some of the essential factors for each organism also change. As these factors change, they become either more nearly optimal or less nearly optimal for the organism. These changes will affect the health not only of an individual, but also of the entire population of like individuals—for example, all the gulls, or all the abalones. Sometimes these changes in the environment are slow, like the coming of an ice age, which takes thousands of years. Sometimes the changes are fast, like a

Figure 2-2 Fresh water runoff from land cuts a route through the beach sand to join the sea. The resulting change in salinity is a major influence on intertidal life.

rain storm that dumps fresh water and silt from runoff on land into a salt water tide pool. Whether slow and unnoticed or rapid and catastrophic, these changes have the same effect: They change the environmental factors. Whatever the change, it will be bad for some living things and good for others.

Because of our dynamic, or ever-changing, environment, living things must also change. They do so in two ways. One way we call *adaptation,* or *evolution,* a slow process that occurs from generation to generation through the process of genetics or heredity. This type of hereditary process allows a

Figure 2-3 This small limpet has two types of barnacles and other limpets on its shell. Finding space to live is perhaps the greatest single competition among marine organisms.

species or population to adjust or evolve to slow changes in its environment. We have long known that a species must adapt to its environment to survive over hundreds or thousands of years. This type of change keeps the organism attuned to the slower changes and, because of its great complexity, is considered to be irreversible. As the organism becomes more adapted to a certain set of ecological factors, it will continue to adapt and become even more suited to that particular environment. Because of this trend to become specifically adapted to a particular set of circumstances, the entire population may become extinct if those circumstances change too rapidly.

The second type of change the ecologists call *succession*. Succession generally happens over a short period of time—sometimes days, sometimes years. The major difference is that a genetically adapting population will survive very slow changes, but in the case of rapid change, it may become extinct. In succession the population is replaced by another population for which the new environmental factors provide the newcomers' optimum living conditions and thus enable them to successfully supplant the original population. If one population were to become extinct and be replaced, the change would normally occur over a large area. Succession, however, also happens in small local areas, such as a bay where humans have changed the environment so that the existing species cannot survive and new ones move in and take over. Another type of succession occurs if a population is killed by some factor, such as a bad storm creating strong waves that knock

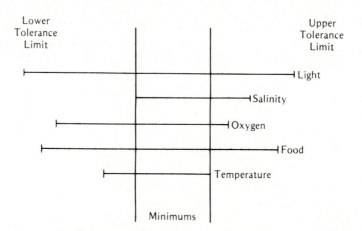

Each line represents the ecological range, or valence, for a particular organisms factors of life. In this hypothetical example, the range is limited by the factors of water salinity and water temperature. The range or distribution of an organism is limited by its factors which show the least tolerance to its environment.

Figure 2-4   An example of life limiting factors for a hypothetical organism.

off all the mussels (a type of shelled animal many species of which attach to rocks), leaving a bare spot. Whatever animal in larval form is in the water at the time will take the spot that was left bare, perhaps barnacles or a number of other types of animals. Even some plants could take over. Thus the living replace the dying.

Succession is normally a very complex process but it may also follow set patterns that help make it predictable. If a piece of wood were placed in salt water off the coast of New York, in a matter of hours it would have a layer of bacteria on it. This would create a slime base, and one-celled animals (**Protozoa**) and one-celled plants (**diatoms**) would adhere to it in the matter of a few days. This base of thicker slime now gives a foothold to animals called hydroids and bryozoans, which by the third or fourth day will be visible through the microscope. By the end of a week or ten days, still larger forms will have settled in the ever-building layer of life. Typical of these are the barnacle (**Balanus**), the mussel (**Mytilus**), and some algae. Because the environmental factors off the coast of New York are particularly favorable for the mussel, by the third month the mussel may well have outcompeted most of the other organisms and be by far the most dominant species on the wood. Because the mussel is a **filter feeder** (filtering the water around it for microscopic food), it will eat the larval forms that try to settle

Figure 2–5 A climax community of mussels (Mytilus) along with barnacles (Pollicipes).

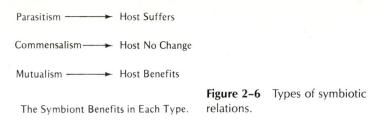

Parasitism ———————▶ Host Suffers

Commensalism ————▶ Host No Change

Mutualism ———————▶ Host Benefits

The Symbiont Benefits in Each Type.

Figure 2–6   Types of symbiotic relations.

near it. This helps the mussels to become a stable population. They will be replaced only when some environmental factor kills them.

Such communities or populations that are the end point of this type of succession and are stable until killed by some major change in the environment are called **climax communities.** Although the process of succession is the same, the kind of climax community will vary according to the type of larva in the water at the time, the type of surface the organism grows on and many other factors. If four different boards were put in the water at the same time, the climax community would be the same on all four. If one

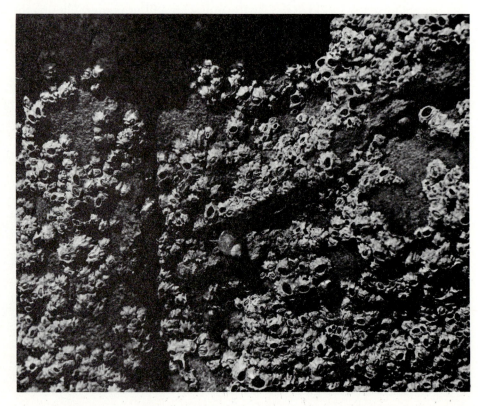

Figure 2–7   A climax community of barnacles. Different climax communities often designate different intertidal zones.

were put in at the beginning of each season, three or four different climax communities would probably result, perhaps mussels, tube worms, tunicates, and barnacles. These changes would be primarily due to the seasonal difference in temperature, light, and the amount of nutrients in the water. Each of these communities would have a select group of other organisms with which it coexisted successfully, but it would be the dominant species. Thus each dominant species along with its group of **symbiotic** species (those that live together) would make up a climax community.

Succession takes place with all living organisms in all environments. We recognize many of the climax communities, such as coral, mangroves, giant kelp, or others mentioned earlier as the major life form in different parts of the world. When you next visit the ocean, be aware of the ever-evolving process from one type of life to the next. Look for areas that have stable communities and compare them with areas that do not appear to be established. Ask yourself why one rock seems to have a stable, abundant population while one close by does not. Although you probably will never know the answer with absolute certainty, an investigation of some of the environmental factors present should give you some good clues. This type of theorizing is what makes marine biology interesting. Let's examine a few of the factors that could be important.

If we concern ourselves just with the seashore, there are a number of

Figure 2-8   The small cleaner shrimp crawl around the gills of some reef fish and clean off small parasites. The fish do not molest the shrimp. It is an example of mutualism.

Figure 2–9   Points of land that project into the ocean like this one are a good example of an unprotected, rocky substrate beach.

obvious things to consider that would affect the type of life that could grow in any small, delineated area. A small area that can be well defined is called a *habitat*. If the area is very small, such as one small crack in a rock or the underside of a rock, we could use the term *microenvironment*.

Figure 2–10   On the 28th parallel in the Sea of Cortez, the rocks are not exposed to constant surf; they are rounded but not worn flat.

Wave impact is a very important factor. Waves possess tremendous physical force which they will exert when encountering an obstacle. If they break on sand, they churn up the sand and move it around. This churning tends to grind up anything living there. Breaking on a hard surface such as rock, they strike and tear loose organisms which are trying to hold on. The wave impact then becomes a limiting factor for any area where it occurs. Some animals and plants have adapted their physical form through many thousands of generations to be able to withstand such conditions. This adaptation allows them to live where most others cannot, thereby reducing the number of competing organisms with which they must live. Typical plants and animals that live on rocks are mussels, barnacles, limpets (many kinds), palm kelp (**Eisenia**), and abalones (**Haliotis**).

A few types that can withstand wave impact in a sandy area are the mole crab (**Emerita,** in some areas called sand crab), pismo clam (**Tivela**), bean clam (**Donax**) and surf grass (**Phyllospadix**). Some corals can also withstand heavy surf and form great barrier reefs with beautiful lagoons behind

Figure 2-11 Eisenia generally lives just below the surf zone but is strong enough to withstand surf action. It can be seen off the coast of Southern California exposed at extreme low tides.

Figure 2-12 Many organisms are territorial; once they have established their territory they will fight to keep off potential competitors. This swimming crab of the Gulf of California is letting us know we are on its piece of beach.

them. If we examine all of the organisms that live in a wave impact area, we would find certain common characteristics. If they are organisms that attach to a solid substrate (such as rock or pilings), they must have a very strong attachment and present a smooth, low resistance area to the water. The **algae** that live in this area bend and flow with the current. If the organisms live in a soft substrate, such as sand, they must be able to burrow for protection and have shells that will withstand the grinding action of the sand grains as they are moved around by the water. Looking even further, we would find that these organisms have developed the need for the high oxygen level found in the foaming aerated water of the surf. Their adaptation to this particular condition is so complete that they cannot survive well in a less harsh area. The old saying about "**survival of the fittest**" means that the organisms most suited to be healthy and reproduce well within any given set of ecological factors are going to outlast or outcompete any other organism less suited to those particular factors. We could say, then, that any climax community is the "fittest" for the set of ecological factors in which it exists.

Another factor we could use to evaluate a habitat is temperature. The deep ocean is the most stable environment in the world. It is so vast and intermixed that sudden changes are totally unexpected; consequently, the animals that live in the ocean, except in tidal areas, do not need wide tolerance to temperature changes or to any other environmental factors. A few

Figure 2-13 When the penshell is dug out of the bottom, its full shape can be seen. Its streamlined form is perfect for burrowing, but its lightly constructed shell needs the protection it gets from being buried in the substrate.

degrees variation of temperature can be quite significant. On land a seasonal temperature change of 38°C (68.4°F) is not uncommon, and a night/day difference of 22°C (39.6°F) is not unusual. In the ocean a seasonal variation of 11°C (19.8°F) is quite high and a daily variation of less than 0.5°C (0.9°F) at the surface is normal. Keeping this stability in mind, we can understand why most subtidal marine animals cannot tolerate a rapid or a substantial change in temperature. There is one habitat that has a substantial temperature change and it fluctuates rapidly and regularly. This is the area of the upper tide pools. When the tide goes out and the trapped water is exposed

Figure 2-14 The high tide pool is possibly the most difficult environment because of the rapid changes in temperature and salinity. Fresh water is a survival problem for tide pool organisms. It changes the salinity, and may dump silt, which clogs an organism's respiratory systems. This is especially true when a heavy rain occurs at low tide.

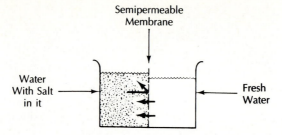

Water will flow into the area of greatest salt concentration through a semiper-meable membrane, lowering the water level on the fresh water side and in-creasing the level on the salt side. Water can pass through the membrane, but salt can not.

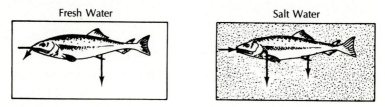

In fresh water the fish must not take in water through the mouth because there is a tendency for the fish to absorb too much water through its semipermeable membranes. It gives off dilute urine to rid its body of extra water.

In salt water there is a tendency for the fish to lose water through its semiper-meable membranes so it takes water in through the mouth, conserves it by passing a concentrated urine, and gets rid of the extra salt it takes in with the water through a salt-removing gland in the gill area.

These two reactions occur because the blood of the fish has salt in it. It is saltier than fresh water, and not so salty as salt water.

Figure 2–15 Effects of osmosis on fish and other organisms.

to the sun, it warms rapidly and can get as high as 38°C (100°F). On cold nights, when the temperature is low, the water can freeze. These extremes make for an environment that is just as harsh as the wave impact area, only more subtle. These conditions can only be tolerated by a few organisms, just as only a few have adapted to the surf. The big difference between the adaptation of the surf organisms and the tide pool organisms is that the tide pool organisms must adapt more physiologically than physically. The ani-mals living in these pools have adapted to wide temperature variation as well as rapid variation. A pool exposed to the sun for a three-hour period and warmed to 21°C (70°F) drops within seconds to 10°C (50°F) when the first wave of the incoming tide floods it.

Similarly, we should consider the salt content of the ocean. Although

Chapter 3 covers this subject in more detail, here let us establish that the salt content, like temperature, does not vary drastically or change rapidly. This fact is important because of **osmosis,** the ability of water to pass through a **semipermeable membrane.** If there is a membrane between two solutions of water which will let water but nothing else pass through, the membrane is *semipermeable.* When less salt is in the solution on one side of the membrane, the water will flow through the membrane to the side that contains the most salt. The greater the difference in the concentration of salt in the two solutions, the more water will flow to the saltier of the two. This phenomenon is important to the animals because their gut and gills are semipermeable membranes, and they will have more or less salt in their blood than the water they live in. Because there is more salt in the ocean water than in their blood, the animals tend to lose some water from their bodies. This can't be allowed to happen because the water is needed to keep the blood from becoming too viscous (thick). Different animals have different physiological methods of coping with the problem. Some have kidneys, others have special salt glands, while others retain urea to make the salt content of the blood about the same as that of the ocean water. Whatever the means of adaptation, any sudden change in salt content of the surrounding water places a strain on the system of the animal. Most animals die. Those that can adapt rapidly change from salt to fresh water in a short time. Salmon (**Oncorhynchus and Salmo**), white-bait (**Clupea**), and the North Atlantic eel (**Anguilla**) are examples. Think of all the tiny creatures that live in the high tide pools. When it rains, fresh water runs into their pool and dilutes it. They must be able to make a physiological shift immediately or they will die. At the same time, the fresh water runoff may carry large amounts of silt from the land into the pools. This silt tends to clog the gills of many gill breathers and in extreme cases will fill a pool and suffocate everything in it.

All things considered, the tide pools, especially the ones that are fairly high on shore, represent the most difficult environment to which marine organisms must adapt. In this higher beach area, marine animals have still another hardship. Twice a day when the tide goes out, they are left high and dry. Some stay in pools of water and suffer the condition we have just discussed, while others seal themselves up and wait for the water to return. Many of this latter group have adapted, throughout generations, and are able to breathe air. A few of them need only be wetted several times a week to live well. Some of them have adapted so well to this area that they must have air for respiration and will drown if they are forced to stay under water for long periods of time. Two of these, the periwinkle and the shore louse, are very common in most parts of the world. The periwinkle (**Littorina**), a very small, shelled animal perhaps only 4 millimeters across, is found in cracks above high tide line on rocky shores. The shore louse (**Ligia**) hides in crevices and is very difficult to catch. Because of their ability to live out of the water 95 percent of the time, they have almost no competition from other marine species for their habitat. It is interesting to speculate how many animals have made that last 5 percent adaptation and migrated completely away from the sea to become land animals.

REVIEW QUESTIONS

1. What purpose does evolution serve for a biological community?
2. How does evolution differ from succession?
3. What is a climax community?
4. How is a rocky environment in the surf different from a sandy one as to living conditions for the organisms living there?
5. Why is the salt content of the water important to the animals?

Chapter Three

A FEW PHYSICAL

AND CHEMICAL

CONSIDERATIONS

DEFINITION OF TERMS USED IN CHAPTER 3

Centrifugal force: The force that pushes something away from the center of rotation.

Coriolis force: A deflecting force caused by the rotation of the earth, which causes the oceans to move in a circular manner.

Drag: A retarding force or resistance against the flow of water created by the friction of the water along the bottom.

EEZ: Exclusive Economic Zone extending 200 miles out to sea from the US coast.

Fetch: The area over which the wind blows to create seas and waves in the ocean.

Neap tides: Neap tides occur between spring tides and have a small range between high and low tide.

Neritic zone: The shallow seas which cover the continental shelves of the world.

Ooze: A soft deposit of mud, shells, and other remains found over most of the ocean floor.

Phytoplankton: Plants which drift in the water.

Spring tides: Ocean tides that occur at full and new moon. The term has nothing to do with season. Spring tides have the greatest range from high tide to low tide in their area.

Thermocline: Layer of water with steep temperature gradient between two layers of water of different temperature.

From Chapter 2 it becomes obvious that in the study of marine biology we must be very much aware of the ecological factors involved in any particular environment we wish to study. Because many of these factors are physical, rather than biological, we must acquaint ourselves with some of the physical factors present in the ocean.

TIDES

Tides have their main influence on the coastal zone and the organisms that live there. The rise and fall of the tidal waters cause very specific tide pool environments to exist along the shorelines. It also helps to shape our coastlines by erosion. Why do we have tides? Ancient peoples noted that the waters rise and fall in relation to the phases of the moon, a relationship so obvious that it has been known for thousands of years. We modern humans find many factors influencing the tides, the main ones being the relative positions of the moon and sun and the centrifugal force.

Being close to earth, the moon exerts a strong gravitational pull and tends to draw the water toward it. It also pulls upon the land masses; but, since the water is more fluid, its bulging toward the gravitational force (see Figure 3–1) accounts for one high tide a day as the earth revolves on its axis. Approximately once a day with each revolution, the moon passes overhead. If the moon were the only influence, each high tide would be just the same

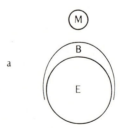

The moon pulls the water ''B'' toward it causing a high tide under the moon at all times.

Figure 3–1   The moon and the tides.

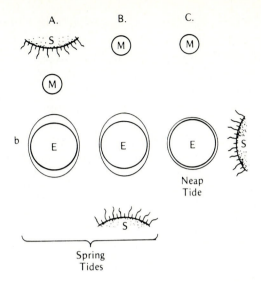

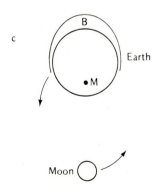

A and B represent spring tides. The sun and moon are lined up and pulling on the same plane. This is a time of higher high tides. C represents a neap tide situation where the sun and moon pull in different directions and tend to reduce the pull in any one direction. The high tides during this phase of the moon are not so high.

The earth and moon spin around their common center of mass, M. Consequently, there is a bulge on the opposite side of the earth from the moon. The water "B" is being pulled away because it is on the outside of a revolving system. This is a similar situation to the water shooting out of a small hole in the bottom of a bucket as it is being swung around in a circle. The water comes out of the hole even when the hole is pointed up.

Figure 3–1 (Continued)

as the last one. Obviously the tides are not the same, and we generally have two high tides a day, not just one. The tides are not the same because of the sun's gravitational pull on the earth, and the second high tide each day is caused by centrifugal force. The sun, although vastly larger than the moon, exerts only about half the gravitational pull of the moon on the earth because it is at a much greater distance from us. Because of the movement through space of the earth around the sun, the moon around the earth, and the earth around its axis, the position of the sun and moon relative to any given spot on earth changes daily. Depending on the position of the sun and the moon, their gravitational forces either pull together to create very high and very low tides (called spring tides, but having nothing to do with the season) or tend to cancel each other, causing neap tides (high tides that are not very high and low tides that are not very low). Spring tides occur, then, at the full moon and the new moon each month, whereas neap tides occur at the quadrature or half moon phase (see Figure 3–1).

Still remaining is the question of why we have two high tides a day instead of just one. As stated previously, this second tide is caused by centrifugal force. Centrifugal force is the force that holds the water in a bucket when the bucket is swung overhead in a large circle. The same type of force that holds the water in the upside down bucket creates a bulge in the water surface of the earth that is similar to the bulge created by the gravitational forces only not so large. This second bulge occurs directly opposite the gravitational bulge of the moon, thus creating a second high tide. The centrifugal force is caused by the moon and the earth acting as a single unit as they travel in their orbit around the sun. The center of mass of the two bodies maintains a straight path around the sun, but the earth is slightly off center

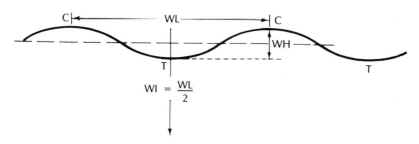

C = Crest of wave
T = Trough of wave
WL = Wave length
WH = Wave height
WI = Approximate depth at which the wave has any major influence on the bottom at sea, equals one half of the WL.

Figure 3–2 Parts of a wave.

Figure 3-3 This pounding surf is characteristic of winter surf over the majority of beaches in the world. The steep beach where the sand has been washed out to sea will disappear when the smaller summer surf returns and moves the sand back up on the beach to create a gently sloping beach characteristic of summer.

in order to balance the mass of the moon; as a result, centrifugal force is created on the earth as it makes its 28-day circle around the center of mass (see Figure 3-1).

All these influences are again modified by the land masses that break up the basic pattern. Each area will have its own peculiarities due to the local topography.

WAVES

Along with the tidal influence on the coast is the influence of the surf. Surf is caused when a wave runs aground on a shallow bottom. Just as a boat would pitch forward if it ran aground and its bottom came to rest, a wave pitches forward and spills over when the wave runs into shallow water and its bottom drags and slows down. We can understand this better by looking at the anatomy of a wave or swell which will become surf when it reaches shallow ground (see Figure 3-2).

A wave has a **crest,** the high point, and a trough, the low point. As the wave moves through the water, the distance between any two like points on any two consecutive waves (crest to crest or trough to trough), is the

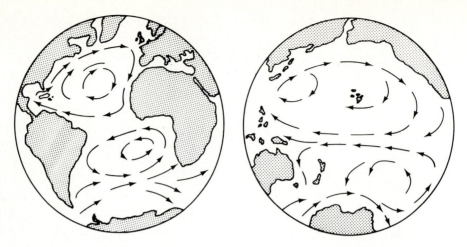

The general circulation of the oceans is clockwise in the northern half, above the equator, and counterclockwise in the southern half, below the equator.

Figure 3–4   General circulation of the oceans.

length of the wave. The *time* it takes any two like points on two consecutive waves to pass a given fixed point is called the **period** of the wave. If we know the distance between two like points on two consecutive waves, we can determine the speed at which the wave is traveling.

We said that surf is a wave that drags on the bottom. How deep is a wave? Where is its "bottom"? The effective depth influence of a wave is approximately one-half of its wave length. The bottom of a wave, then, is as deep as one-half of its wave length. If a wave has a length of 100 meters, its bottom would be at 50 meters. If the water is deeper than half the wave length, the wave is called a *deep water wave*. If the ratio of water depth to wave length becomes as low as ¹⁄₂₅ or if the wave with a length of 100 meters gets into water as shallow as 4 meters (⁴⁄₁₀₀ = ¹⁄₂₅), we call it a *shallow water wave*. In depths between 50 meters and 4 meters, it could be called an *intermediate wave*. Intermediate waves "feel" the bottom and may change shape and direction because of it, but probably will not break to cause surf until the wave becomes a shallow water wave. Waves generally will break when the water depth is approximately 1.3 times the wave height. Thus a 3-meter wave will break when it reaches an approximate depth of 4-meters.

(depth × wave height = ⁴⁄₃ = 1.3)

The shape of the surf depends on the shape of the bottom. If the bottom slopes gently, the waves will break softly and roll to shore. This wave is called a **spilling breaker.** If the beach is steep and the wave comes on it suddenly, it will hump up and crash very abruptly in a **plunging breaker.** If the bottom is so steep that the wave does not have time to react, it will not break at all, but will instead push water up on land and suck it back out. This type of wave, called a ***surging breaker,*** normally occurs on sides of cliffs or artificial jetties. The spilling breakers are perfect for swimmers to play in and surfers to surf on. Plunging breakers are dangerous for swimmers, and only the most advanced surfers can ride the tube which is formed as they plunge over. The surging breakers are encountered by anglers and divers who frequent the steep rocky slopes. They are sometimes caught by surprise and swept off the rocks. Small children are particularly susceptible to this danger.

There are other types of waves of which we should be aware; however, only a brief description will be given here. **Tsunamis** are seismic sea waves, called "tidal waves" by the layperson. They have nothing to do with the tide any more than a starfish is a fish. They are generally caused by large shifts in the crust of the earth along some fairly shallow area (46 meters or less). The shift must be large and along many miles (60 or more) of coastline. Other causes of tsunamis are underwater landslides and explosions. All of these causes are generally noted at the onset of a strong earthquake. Because these waves can cause great destruction, a warning system is now in effect, and areas that might be subject to damage are told of the oncoming wave in advance. A few hours' warning can save many lives.

Storm Waves

Storm waves are not so much waves as they are banks of water piled up by high winds. The wind blowing strongly against a shore will pile water higher than normal on that shore. If this happens during a spring high tide, the increase in water height can be as great as in the case of a tsunami. Storm waves generally occur during hurricanes.

Origin of Normal Sea Waves

Where do normal waves come from? Most waves that travel across the seas of the world are caused by the wind, which makes ripples on the surface of the water. If the wind stops blowing while there are still only ripples on the water, the sea will become flat again; but if the wind blows long enough and strong enough, the ripples get larger and continue after the wind stops. The area where the wind is blowing and creating waves is called the **fetch.** The waves *in the area of the fetch* are called **seas.** After they leave the fetch area, they smooth out and are called **swells.** Seas are steep-sided and rough for boats, whereas swells are smoother and generally not so uncomfortable for boating. If the winds blow in a fetch area strongly enough, and for a long time, the resulting swells can be very large, as high as 30 meters. The normal swell is around 3 to 5 meters. The swells from one fetch will be nearly the same in size and will travel out across the ocean until they are stopped by some land mass. Swells from different fetches, which often are crossing the ocean at the same time and from different directions, cause a very irregular or rough sea. Occasionally swells from one area will catch and coincide with swells from a different area so that the two swells join and create a single, very large swell. Clearly, the surface of the sea is extremely unpredictable and interesting.

CURRENTS

Everyone understands that a current is water moving from one place to another, but not everyone understands the great significance of this movement. Without currents, the oceans would become stagnant and would support far less life than they do now. Food, nutrients, and oxygen are three of the main substances that must be circulated throughout the oceans for life to be widely distributed. These materials typically are produced or enter the sea in prescribed areas or zones and, through circulation of the water, are carried to all parts of all seas to support life.

In normal scientific usage, a **current** is water moving one knot or faster. Water moving more slowly than that is often referred to as a **drift.** For example, the slow-moving water around the Antarctic continent is called the West Wind Drift, whereas the fast-moving water off the east coast of North America is called the Gulf Stream Current.

GENERAL CIRCULATION OF OCEANIC WATERS

The general patterns of circulation in the major ocean basins are, in the main, controlled by three factors: wind, Coriolis force, and drag created by the rotation of the earth. Wind is perhaps the most significant of the three, as it has a positive effect at almost all points. The general wind circulation, for reasons beyond the scope of this book, is to the west along the equator and to the east near the poles. These winds then push the surface water along in the same direction. Because there are large land masses (continents) to interrupt this pattern, the water is deflected as it moves with the wind. We find that the water in the Northern Hemisphere moves to the right, whereas the water in the Southern Hemisphere moves to the left as it runs aground on the east coast of the continents. The reason for this right (north) and this left (south) direction is the **Coriolis force.**

The *Coriolis force* is an abstract concept. It is caused by the shape of the earth and the fact that the earth spins on its axis. The speed at which a point on the surface of the earth is traveling depends on its distance from the poles. For example, in approximately 24 hours, the earth turns around its axis once. If you are standing on the equator, in order to travel around and come back to the same place, you must travel many thousands of miles. If you are standing one mile from the pole, you will only travel a few miles during the same 24-hour period. Visualize, for example, the speed at the center of a phonograph record (the pole) in relation to the speed at the outer edge of the record (the equator). If you rolled a marble from the center of a record to the outer edge while the record was turning, you would see the marble curve away from the direction in which the record was turning. The ocean currents do the same thing. They curve away from the direction the earth is turning and move toward the slower-moving portion of the earth. Thus the currents along the equator are deflected toward the poles. This Coriolis force, then, adds to the velocity of the current as it travels north or south along the eastern edges of continents; but along the western edges of continents, the Coriolis force acts to slow the current. Consequently, we have our faster currents on the eastern shores of continents, like the Gulf Stream on the east coast of North America and the Japanese Current (Kuroshio) on the east coast of Asia.

Drag is a negative force that tends to slow the currents. The faster you ride a bicycle, the more drag you experience from the air. If you ride slowly, you feel almost no wind resistance (drag). If you ride faster, the drag from the air becomes evident. The drag created by the sea floor is similar: the faster the current flows, the more drag it encounters. This is why currents move faster, in general, at the surface than they do near the bottom. The interaction of these three forces is responsible for the fact that the currents in the Northern Hemisphere, in both the Atlantic and the Pacific oceans, rotate clockwise and the currents in the Southern Hemisphere rotate counterclockwise.

There are deep water currents that we will not discuss at any length. In general, they run in the direction opposite to the surface currents and seem to replace water within an area where it has been swept away by surface currents. The main counter-currents are deep along either side of the equator, and the Gulf Stream counter-current which runs from the Nova Scotia area to the tropics beneath or in the general area of the Gulf Stream. The exact paths of all currents fluctuate due to seasonal and other influences.

Local Currents

Every area has its own local currents. Caused by factors such as wind, land forms, and tides, these local currents are important to boaters, anglers, divers, and swimmers. Most shore areas have alongshore currents a great deal of the time. These currents are caused by the direction of the swells as they come onto land. In general, the swell does not approach land at a 90° angle; it normally comes from either the left or the right as we look out to sea. Consequently, the general water flow in the surf line is up or down the beach. In heavy surf conditions, this can be a strong and dangerous current. Normally it is a slow drift. The sand on the beaches is moved along with the current, and it is this sand movement that carries the sand from the mouth of rivers along the coastlines and forms beaches. As the sand is moved along, it can be lost into submarine canyons if the canyons are close to shore. Along the coast of California at Monterey and La Jolla are two major canyons which trap much of the sand. The sand enters the canyon and is swept down to the ocean floor. Another famous place where this occurs is off Cabo San Lucas in Baja California, Mexico. One can dive to a depth of 45 meters and watch the sand tumbling over the rocks, much like a waterfall on land.

Tidal Currents

Tidal currents are caused by the rise and fall of the normal daily tidal changes. On a beach open to the sea, there will be no tidal currents; however, if the water must go through a passage or around an island, currents will be formed. Tidal currents are characterized by flow in one direction as the tide comes in, no flow at all at peak high or low tide (called slack water), and flow in the other direction as the tide goes out. Examples of areas where the land formation is such that strong tidal currents occur are British Columbia, Nova Scotia, San Francisco, Gulf of California (Sea of Cortez), Chesapeake Bay—in fact, any bay area where the water flow is restricted.

If the currents are fast flowing, they are responsible for cutting channels in the ocean floor at the mouth of harbors and bays. If they are slow currents, they tend to let material drop out and "silt up" a channel. During spring tides, the currents run faster because of the greater differential between high and low tide levels. Often where humans have not modified the area, a sand bar will appear just outside the channel of a bay because, as the fast current through the channel slows in the open sea, the sand it is carrying drops out. Along the Pacific coast of Baja California, Mexico, these sand bars are common because the coastline is unmodified by humans.

Upwelling

Upwelling is a process where deeper water is brought to the surface. This condition can occur when a wind blows the surface water out to sea and the deeper water moves up to replace it. This can be a local, temporary situation or cover large areas and be fairly permanent.

Some of the major upwellings that occur constantly in the same areas

Figure 3–5 One method of sampling the fish in an area is to construct traps. Fish traps work well in areas where nets cannot be used, such as in kelp beds, under ice, or in rocky areas.

are caused by the deep currents running into continental land shelves. Good fisheries develop in these areas because of the high nutrient content of the deep water that has been brought to the surface. The plant plankton (**Phytoplankton**) grow rapidly as any plant would in a fertilized field. Such areas as South Africa, Peru, and Equador are examples.

In general, when any substance is heated, it expands; when it cools, it contracts. Although water also expands and contracts, it deviates in one very important way from the normal reaction of other substances. Most things continue to contract (decrease in volume) the colder they get. Water does

Off the Southern California coast upwellings occur whenever there is a Santa Ana wind (a warm local wind that blows from the shore out to sea). The local sea temperatures will change as much as 5°C (9°F) in a matter of hours as the colder deeper water moves in. This condition rarely lasts more than a few days. Off the west coast of Peru, the prevalent southerly and southeast winds, along with the deep currents, cause a strong upwelling carrying the surface water away from the coast. Because the deeper water is richer in nutrients, a heavy plankton bloom is associated with most upwellings. When this becomes a steady condition, as it has off the coast of Peru, the fish move into the area permanently to feed on plankton. As would be expected, Peru has one of the world's best fishing grounds just off its coast. Many countries that have good fishing grounds off their coast, because of some phenomenon such as the Peruvian upwelling, try to keep control over the fishing grounds by passing laws to extend their boundaries 800 to 1,300 kilometers (approximately 500 to 800 miles) out to sea. Legislation of this sort is causing international political conflicts.

Temperature

The various changes in physical characteristics of water that occur as the result of temperature changes are both interesting and important.

In general, when any substance is heated, it expands; when it cools, it contracts. Although water also expands and contracts, it deviates in one very important way from the normal reaction of other substances. Most things continue to contract (decrease in volume) the colder they get. Water does not. Water contracts and gets heavier (more dense) only to the temperature of 4°C (39.2°F); then it starts to expand again as the temperature is lowered further. When it freezes, the ice is less dense than the water and floats on the surface. Imagine what would happen if water, like other substances, continued to contract as it got colder. As the water cooled to the point of freezing and formed ice, the ice would be more dense than the water and would sink. The cold polar air would freeze the surface water and it would sink. More ice would form and sink. Soon our entire ocean would cool and start to make the air even colder; in turn more ice would form, and we would turn into an icebound planet. Instead of this happening, as the air cools the ocean at the poles and the water temperature reaches 4°C (the saltier the water is, the lower the maximum-density temperature is), the oxygen-rich surface water sinks to the bottom of the sea and mixes with the deep water of the ocean basins.

In the tropics the more direct rays of the sun heat a layer of warm water on the surface. This warm, less dense water tends not to sink, so a layering or stratifying of the oceanic waters occurs. These various layers or strata can be identified by characteristics such as temperature. For example, the cold water that sank in the Antarctic is pushed by more cold sinking water along the ocean floor until it covers most of the ocean bottom. Very deep bottom samples of the water in the tropics show water very similar to the surface water in the Antarctic. This subtle circulation pattern is one of the important circulation systems in the ocean. The surface water slowly migrates toward the poles, and the bottom water slowly migrates toward the equator. One of the more interesting results of this stratification of the water is the clear water associated with the warm tropical surface water. This warm water is less dense and stays on the surface. The plankton in it use up the nutrients, and it becomes a planktonic desert with a relatively small number of plank-tonic forms surviving. Because of the small amount of plankton in the water, it is much clearer than the richer temperate zone areas that support more life.

Thermoclines

The word **thermocline** means *temperature gradient*. The term is also used to describe the transition layer where two strata of different temperatures meet. Some thermoclines are layers of very rapid temperature changes, such as in the submarine canyons off La Jolla, California, where a change of 5°C (9°F) in 3 meters is not uncommon at depths of about 30 meters or off Palancar Reef in Mexico where at approximately 45 meters the temperature will change suddenly about 1°C (2°F). Other thermoclines take place over a larger range and are permanent. Temperatures generally drop from the surface to the bottom in all areas except perhaps the polar seas, where the water is uniformly cold. In summer the surface waters are warmed by the sun to a greater extent than in the winter, so the thermoclines become more apparent and more sharply defined. The water column becomes stratified.

Density

Water, like all materials, is composed of molecules. The closer these molecules are to each other, the more dense the object is. Steel has more closely packed molecules than does water; therefore, it is more dense (heavier). We have already said that water becomes less dense as we heat it, as do other materials.

Let us draw an analogy between the molecules of a substance and people on a dance floor. If we have 100 molecules in a substance and 100 couples on a dance floor, with the music representing the energy given to the molecules by temperature, it is easy to see what happens. When a slow number is played, the couples move slowly around the dance floor, and all 100 couples can dance in a small area. When the tempo of the music is speeded up, the couples move faster and need more room; consequently, the dance floor must be enlarged for them all to dance at the same time. The molecules react in the same way to increasing heat. They become more active and need more space; consequently, the substance expands. Water, remember, contracts as the temperature gets colder until 4°C, and then ex-

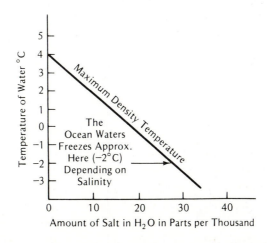

Fresh water is most dense at 4°C. The saltier the water the lower the maximum density temperature. In the oceans it is in the area of −2°C.

Figure 3–6 Maximum density of water.

pands slightly and stays stable. Temperature, then, directly affects the density of the water because of the reaction of its molecules to heat energy.

Other factors, mainly salinity and pressure, also affect density. Pressure influences density in a fairly uniform manner worldwide. The pressure increases with depth, as we would expect because of the weight of the water above any given depth. The weight or pressure exerted by the water above any given point is equal to one atmosphere 1,033 g/cm^2, (14.7 pounds per square inch) for every 10 meters. Water is not very compressible; therefore, the density change caused by pressure is not very dramatic.

Salinity causes more significant changes in density than does pressure. The saltier the water, the lower the maximum temperature. As a result, the cold water on the bottom of the deep ocean is not 4°C (39.2°F) as it would be in a deep fresh water lake, but is more like −2°C (28°F) because of salinity.

Many salts are dissolved to make up the salts of sea water (see Table 3–1). Although the total salinity varies from place to place because of evaporation and the addition of fresh water from various sources, the ratios of one salt to another remain identical in all areas. To determine the density of a water sample, the scientist measures the salinity by one of several methods, taking into account the temperature and the depth at which the sample was taken. Because the ratio between salts is constant, any one of them could be measured and the others arrived at by mathematical deduction. Normally we measure the amount of chloride.

Density affects organisms such as plankton because plankton have nearly the same density as the water mass in which they live. A cold, denser layer of water under a warm layer will stop much of the dead plankton from sinking to the bottom and hold it at the thermocline. This decomposing organic matter will use up the oxygen and create a zone of low oxygen.

TABLE 3–1 SALTS DISSOLVED IN SEA WATER IN ORDER OF ABUNDANCE

Ion	Approximate percentages per weight
Chloride—Cl^-	55.25
Sodium—Na^+	30.50
Sulfate—SO_4^{--}	7.70
Magnesium—Mg^{++}	3.75
Calcium—Ca^{++}	1.22
Potassium—K^+	1.11
Bicarbonate—HCO_3^-	0.21
Bromine—Br^-	0.19

Figure 3-7 When water samples are brought back on deck, the data are carefully recorded so the exact depth and temperature are known. The man on the right is reading the temperatures from the reversing thermometers attached to the Nansen bottles. His companion is recording his observations.

THE OCEAN FLOOR

Only in the last 40 years or so have we had the technology to study the ocean floor with any degree of accuracy. In 1949 Ewing and Press established that the earth's crust beneath the ocean floor and the earth's crust beneath the continents had important differences in composition. This major discovery raised many questions but gave very few answers. Scientists are now hard at work to help fill this gap in our knowledge. Some of the facts we have established have commercial, as well as scientific, applications, such as in oil exploration.

The theory of **plate tectonics,** developed in the late 1960s, has increased our understanding of the movement of the earth's surface as well as the bottom of the deep oceans. This theory has been modified with new

Figure 3–8 The Nansen reversing water bottles are attached to a metal cable and sent to whatever depth is to be tested. Many can be placed on one cable to sample various depths at the same time. A weight then sent down the line "trips" them so they reverse and take a sample at their particular depth.

findings in the late 1980s that show that the plates are thicker than previously believed and extend down into the **mantle** of the earth. Because the mantle makes up 83 percent of the earth, our new understanding of it is of great importance. **Geophysicists,** who study such information, are on the leading edge of finding new information about our planet.

Although the average height above sea level for the land masses of the world is approximately 837 meters (2,755 feet), the average depth below sea level is slightly over 3,798 meters (12,500 feet). It is this great depth which has kept the floor of the sea out of reach for so long. New technology has not only developed many remote sampling techniques, but also submersibles in which humans can penetrate the deepest parts of the ocean (over 10,937 meters) and observe what is there. The information being accumulated by hundreds of researchers and their modern technology is being fed into computers and new correlations are giving better insights into the dynamics of the oceans.

The Continental Shelf

A shelf of land gently slopes from the surf line to a depth of approximately 200 meters (600 feet) off most of the continents of the world. While it is very extensive off the coast of Siberia (over 1,400 kilometers), it is nonexistent off the coast of Chile. Because this area of ocean floor is accessible, it is very important to humans. The continental shelf is close enough for land-based fisheries, shallow enough for oil drilling, and its waters are well mixed with oxygen and nutrients which support dense populations of living organisms.

The shelves of the world appear to have been created by a combination of wave action during glacial periods, when the ocean level was lower, glacial scouring and deposits, silting from water runoff and the sinking of land masses. Because of its complex makeup, the continental shelf holds great interest for the submarine geologist.

The water that covers the shelf, because it is normally exposed to sunlight and rich in nutrients from the upwelling of deep water over its edge

Figure 3–9 Scientific divers often use small inflatable boats to get to areas where they can record observations and make plankton tows.

and runoff from land, supports a rich planktonic population. Fish come in to feed on the plankton and create a good commercial fishery. Many of the plankton found in this zone are called **meroplankton.** This term designates planktonic forms that do not spend their entire life living in the plankton, such as some larval forms of crabs and fish, and distinguishes them from those that exist only as planktonic forms (**holoplankton**). The general area over the shelf is referred to as the *neritic zone,* and the organisms that live there are called *neritic species.*

The continental shelf generally ends rather abruptly at what is called the continental edge, where the bottom drops off more rapidly. This dropoff is called the continental slope. The slope normally continues down to the deep ocean floor; however, at some places along the slope, interruptions exist. These phenomena, depending on their characteristics, could be a submarine canyon such as the Hudson or Monterey canyons, or a sea mount rising from the sea floor.

New methods of mapping the ocean floor use sonar. Sonar maps are so detailed that they have been kept classified by the government. The reason for the security is that foreign submarines could navigate along our coast without ever coming up to see where they were. The area mapped comprises 3.9 billion acres in an area known as United States **Exclusive Economic Zone** (EEZ). This is an area extending 200 miles off the coastlines of the US.

The Deep Ocean Floor

As one would expect, the ocean basins are covered with sediments. These sediments vary in different sectors of the sea, but there is a general pattern to their distribution. Around the continents we find a very heterogeneous

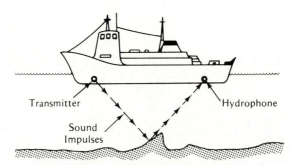

Echo sounding is one of the ways that the depth of the ocean bottom can be charted. The time it takes the impulse to travel through the water gives an accurate depth.

Figure 3–10   Finding the depth of the ocean bottom.

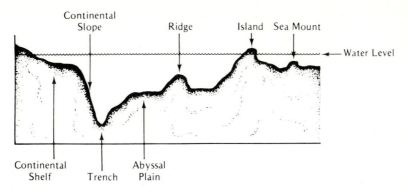

Figure 3-11   Bottom features with considerable vertical exaggerations.

or mixed group of sediments composed of terrestrial materials, biological remains, and, in polar regions, glacial sediments.

In the centers of most of the large ocean basins are **red clay** deposits. **Siliceous ooze,** the name given to the deposits mainly left by plants called diatoms and one-celled animals called radiolaria, is found in great abundance along the 60th parallel south around the Antarctic, in smaller amounts along the equator in all oceans, and at 55° north latitude in the Pacific basin.

By far the largest area of ocean bottom is covered by **calcareous ooze.** This type of sediment is composed mainly of foraminifera and pteropods, animals that leave shells containing calcium. This type of deposit is not common below 5,500 meters because the calcium will dissolve under the pressure and temperature below that depth. The deepest parts of the ocean are typically red clay, with particles of siliceous ooze occurring in all oceans but to a greater extent in the Pacific, where there are large deposits in the north, on the equator, and in the south.

The rate at which sediments are deposited varies from season to season, year to year, and place to place. There is no standard rate of deposition.

Figure 3-12   Most of the many types of bottom samplers are sent to the bottom in an "open" position and close when they strike the bottom. This allows the scientist to bring up a small sample of the bottom and discover its composition. By taking many samples over a large area, a "map" of the benthic substrate can be constructed.

TABLE 3–2 COVERAGE OF OCEAN BASINS

Material	Percentage
Calcareous ooze	48%
Red Clay	38%
Siliceous ooze	14%

Among the interesting items found on the sea floor are manganese nodules. The metal found in the greatest amount is manganese, along with nickel, cobalt, and others. This resource may some day be mined, for it has great economic potential.

The deeper one samples, the fewer types of life forms are found. A bottom sample at 1,000 meters would be expected to turn up 100 to 200 species, whereas a sample at 7,500 meters would be expected to turn up 5 to 15 species. The mineral content of the sediments is not so important to life as the organic or nutritional content. Shallow seas generally will have sediments containing a higher proportion of organic materials.

The discovery of deep hydrothermal vents that support life based on hydrogen sulfide revealed entirely new biological communities on the deep ocean floor that were completely unexpected. In the areas where these hydrothermal vents occur, life abounds. This life is not based on light as other life on earth is, but rather on a chemical compound called *hydrogen sulfide*. This discovery creates a new frontier in deep oceanic biology.

REVIEW QUESTIONS

1. What areas are most influenced by the tides? Why?
2. What is surf, and how is it created?
3. What are the factors that create the general ocean current pattern?
4. What is the relationship between tidal currents and the formation of sand bars?
5. What does water density have to do with the oxygen minimum layer?
6. Why is the EEZ important?

Chapter Four

ENERGY AS FUEL FOR LIFE

DEFINITION OF TERMS USED IN CHAPTER 4

Autotrophic: An organism that through photo- or chemosynthesis produces its own nutrition: it is a primary producer.

Chemosynthesis: Formation of carbohydrates through the use of chemical reactions instead of light.

Detritus: Material forming a scum on the surface of ooze and mud environments; composed mainly of decaying plant material and the bacteria breaking it down.

Euphotic zone: The top layer of the ocean where there is sufficient light energy for photosynthesis. The depth varies with the clarity of the water. In clear tropical water, it is almost 200 meters deep.

Heterotrophic: Any organism that is not autotropic: a secondary producer. Utilizes the organic material produced by an autotrophic organism.

Pelagic: Referring to any organism that spends its life in the open sea away from the bottom, such as the tuna.

Photosynthesis: Formation of carbohydrates through the use of light.

Primary producers: Organisms that use chemical energy or sunlight and the process of photo-or chemosynthesis to produce organic material.

All living things need fuel or energy to run the process we call life. This fuel takes many forms, and different organisms need different forms to survive. The basic component of most forms of energy is sunlight.

In 1977, life forms were found that used **hydrogen sulfide** as their base to form energy, and in 1986 life forms that use **methane** were isolated. These organisms have created an entirely new field of biology. Heretofore, we had believed that all primary producers were plants that used sunlight, along with the process of **photosynthesis,** to create usable energy for animal life. Now we know that some life forms are chemically dependent and use an entirely different process called **chemosynthesis** to create usable energy for higher life forms. This chemosynthesis is carried out by bacteria, which multiply and act as food for higher life forms. This discovery creates many questions and many possibilities. One such question is, "Can we use this process to get rid of hydrogen sulfide, which is produced at our waste-disposal plants, and produce food for humans at the same time?" Some scientists at the Woods Hole Oceanographic Institution in Massachusetts think we can. The discovery of chemosynthetic life also opens up the possibility of life on other planets where we thought there could be no life because of the particular atmosphere on them. Discovery of these new life forms has been a very exciting breakthrough for the scientific community.

The organisms that use sunlight directly and convert it to stored energy are also *primary producers.* This process is generally carried on by plants, which combine inorganic materials into organic materials, such as carbohydrate and protein, with the aid of sunlight. All of the food on earth comes from the primary producers. The basic materials produced by the plants are carbohydrates. These carbohydrates are then converted into the other materials needed, such as fats and oils, starches, proteins, and other sugars. The majority of life in the ocean, like life on land, is totally dependent upon sunlight to survive.

Although the euphotic zone may extend to almost 200 meters in the tropical areas, in the temperate zones it is rarely deeper than 80 meters and is usually considerably less. All of the photosynthetic primary producers must live in the euphotic zone; however, many of the nutrients needed for plants to produce food are below the euphotic zone. Consequently, a mixing of the deeper water with the shallower waters is necessary for the euphotic zone plants to produce. This mixing takes place in many ways.

$$12\ H_2O\ +\quad 6\ CO_2\quad +\ Light\ Energy\quad \text{(Chlorophyll)}\qquad C_6H_{12}O_6\quad +\quad 6\ O_2\ +\ 6\ H_2O$$

Water + Carbon Dioxide + Light Energy $\underrightarrow{\text{(Photosynthesis)}}$ Sugars and Starches + Oxygen + Water

This very simplified diagram shows the basic food production of all plants. This type of food production is the basis of most food production, both on land and in the sea. In the sea it is accomplished by the phytoplankton and algae.

$$6\ CO_2\quad +\quad 6\ O_2\ +\quad 24\ H_2S\quad +\ Bacteria\ \xrightarrow{\text{(Chemosynthesis)}}\quad C_6\ H_{12}\ O_6\quad +\ 24\ S\ +\ 18\ H_2O$$

Carbon Dioxide + Oxygen + Hydrogen Sulfide + Bacteria $\longrightarrow$ Sugars and Starches + Sulfur + Water

This process is not yet fully understood, but seems to work approximately in the manner outlined above.

Figure 4–1   Basic food production.

Winter mixing of the water is far more extensive than summer mixing. In winter the surface waters are not warmed as much as they are in the summer, when the period of daylight is longer and the rays from the sun more direct. Because the surface waters remain cooler, there is less difference in temperatures between the bottom and the surface. This makes the density of the waters more nearly equal, and thus allows the deep water to rise to the surface more easily. Wind and currents keep the ocean fairly well mixed if there is little difference in density. In the summer when the surface water is warmer, the deep water has difficulty rising because it is colder and denser than the warm surface water. The plants in the upper layer of water use up the available nutrients and stop production. In the tropics, where this layer of warm surface water is fairly stable, it lacks the nutrients to support much plankton, so the water is generally clear. The clearer the water, the more of a desert the area is. The clear, warm water of the tropics is great for scuba diving and recreation, but very poor for commercial fishing. The good commercial fisheries are located where the surface water allows for nutrient mixing, in the mid-latitudes.

The amount of organic material that is created in the photosynthetic zone is the total or gross productivity of photosynthesis in a given area. Plants, like animals, are alive and through their metabolism or life processes use up some of the organic materials or energy they create. The **net productivity** is the difference between the gross productivity and the amount of organic matter utilized for metabolic processes. The net productivity is what is important to the animal life of the area. The majority of organisms live at the surface or near the bottom due to food gathering. It is more difficult to gather food at mid depth than it is near the surface, where it is produced, or on the bottom, where it tends to collect.

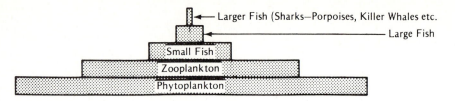

← Larger Fish (Sharks—Porpoises, Killer Whales etc.

—— Large Fish

Small Fish

Zooplankton

Phytoplankton

Figure 4–2   Pyramid showing inverse ratio of fish size to fish population. There are few large fish, there are lots of little fish, but most numerous are the minute organisms.

The majority of sea plants are members of the phytoplankton. This means that they must exist in the top layer of the sea to receive enough sunlight to survive. It has been estimated that approximately 20 percent of this phytoplankton sinks below the required light level and dies. That 20 percent accounts for about 90 percent of the energy available to the deep water animals. The rest of the energy comes from *dead zooplankton* (animal plankton) or dead *pelagic predators*. This dead type of food is called **Detritus.** The transfer of energy from the sun to the primary producers, to the

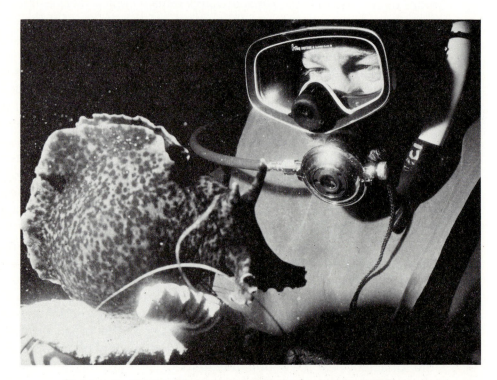

Figure 4–3   The "sea hare" observed here by a diver is capable of producing 86,000,000 eggs a year. All but a few of these become part of the food cycle of other marine organisms. (Photo by Dick Clucas)

benthic animals through the detritus system, is the major energy transfer to deep water for living organisms. In shallower waters such as on a continental shelf, there is organic matter washed out from land, which includes wood, leaves, sewer outfalls, etc., and is about equal to the natural detritus available. In either case, whether the available detritus is from shore runoff, made by humans, or sinking phytoplankton, only about 25 percent of it is used directly as food by the benthic plant eating organism; the other 75 percent of the food source for these benthic types is the bacteria that flourish on the detritus. Various food sources, such as bacteria that form on organic substances but are not photosynthetic, are called *secondary producers.*

Those bacteria that derive energy directly from hydrogen sulfide and do not require sunlight are primary producers. Although these chemosynthetic bacteria normally occur in the deep ocean around warm water vents they also can occur in shallow water as they do at White's Point in Southern California.

To account for total gross or net production, one must consider all forms of production and their use.

The closer one gets to the primary producers in this energy transfer

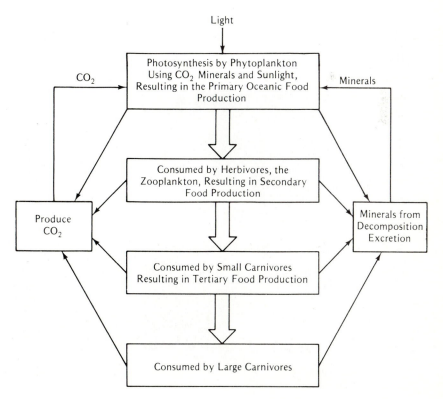

Figure 4–4   A simple photosynthesis energy transfer chart.

chain, the more efficiency one finds. For instance, production efficiency in the phytoplankton is approximately 0.1 percent. Fish, which are generally several steps removed from these primary producers, have an efficiency level of around 0.0001 percent. As a food source, then, phytoplankton would be much better than fish, as far as utilization of food produced is concerned.

The production of any given area depends upon many variables; thus it is always an estimate rather than an exact figure. Even estimating is difficult because the numbers from one area, although accurate for that area at that specific time, are not transferable to any other area with any degree of accuracy. This is unusual for science. Generally, science solves its problems by creating a controlled situation and conducting an experiment. The results of the experiment are then transferred to like situations. Although this method works very well for most things, it does not work well for productivity predictions because too many variables are involved. Productivity at any given point changes with intensity of sunlight, temperature, seasons, availability of nutrients, and even the type of organisms. Organisms can either create food through chemosynthesis and photosynthesis or ingest food produced by another organism. The terms for organisms at different levels of this energy transfer chain are identified by the suffix **trophic,** which means nutrition. Primary producers that create their own food are **autotrophic.** Plants and animals that ''eat'' organic substances created by the autotrophic organisms are **heterotrophic.**

Figure 4–5 The squid is often used for food and for bait to catch fish on hook and line. Thus it is part of the food chain for humans as well as other marine organisms. Like the octopus, it can swim by jet action. They live in schools and feed on small fish.

Figure 4–6 Schools of small fish eat zooplankton; in turn, they are food for larger fish.

Through extensive research involving the cooperation of many nations, we have enough data to generalize about the productivity of the sea. Although estimates may be inaccurate, perhaps even by a factor of 10, we are close enough to obtain the necessary data needed for recommendations to improve the environment, to benefit commercial fisheries, to minimize pollution, and to increase food sources for the expanding world population.

Energy use by the environment is always as complete as possible. Whenever energy is wasted in a system, something will either move in or an existing organism will become modified to use the energy not being consumed. Consequently, most environments have very specific types of organisms closely associated with them because of the dependency on a specific food or energy source. The longer the environment has been stable and the less variation that occurs in it from season to season, the more specific types of organisms will occur there. The tropics are the best example of this phenomenon. Many varied types of living creatures exist there, each closely dependent on the other and a nonchanging environment for survival. The energy available in the tropics is very efficiently used by these specialized organisms.

REVIEW QUESTIONS

1. Why is the sun so important to the life in the sea?
2. Why do the waters mix better in the winter than in the summer?
3. Why would plankton be a better human food source than fish?
4. Why is it difficult to determine the productivity of the ocean?
5. Why do the tropical regions have so many organisms compared to other regions?
6. What is the difference between photosynthesis and chemosynthesis?

Chapter Five

UNITS OF LIFE

DEFINITION OF TERMS USED IN CHAPTER 5

Cell: Living material contained by a cell membrane. Considered the "basic" unit of life (cell theory).

Cell membrane: Semipermeable membrane consisting of three layers surrounding a cell.

Cell wall: A nonliving ridged structure composed mainly of cellulose, which gives form to plant cells.

Cellulose: An insoluble complex carbohydrate.

Centrioles: Found in Eukaryotic animal cells, as well as the sperm cells of some plants. Function during cell division to form asters.

Chromatid: One-half of a duplicated chromosome.

Chromosome: Found in the nucleus; contains DNA, and controls inherited traits.

Cytokinesis: The division of the cytoplasm.

Cytologist: One who studies cells.

Cytoplasm: Living matter of a cell excluding the nucleus.

Diploid: Having complete pairs of chromosomes. 2N.

DNA (Deoxyribonucleic acid): Carrier of genetic information in cells.

Endoplasmic reticulum: A membrane system dividing cellular cytoplasm into compartments.

Eukaryotic cell: Contains a nucleus.

Gametes: Sex cell; sperm, egg, or spore; haploid cell, formed by gametogenesis; meiosis.

Genes: A unit of heredity in the chromosome.

Germ cell: Sex cells, gametes; formed by meiosis.

Golgi bodies: Storage area within a cell.

Gonads: Organ in which meiosis takes place.

Gross anatomy: The study of complete organ systems.

Haploid: Chromosomes that are single not paired. 1N.

Histology: The study of tissues.

Lipid: Organic fats and oils.

Meiosis: Process used to form sex or germ cells; sperm, egg, and spores.

Mitochondria: Produce enzymes to convert food to energy.

Mitosis: Process of producing diploid or somatic cells.

Nucleus: A membrane separated part of a eukaryotic cell containing the chromosomes and the nucleolus. Generally the most prominent feature in the cell.

Oogenesis: The process that forms the egg.

Prokaryotic cell: Primitive cell type found in bacteria and the blue-green algae. These forms of life are classified as Monera.

Protoplasm: Umbrella term for all living cellular substances.

Ribosomes: The site of cellular protein synthesis.

RNA (Ribonucleic acid): Works in conjunction with DNA to carry out heredity orders to the body.

Somatic cells: All cells of an organism except the germ cells.

Spermatogenesis: Formation of sperm; meiosis.

Vacuoles: A space in the cytoplasm.

Zygote: A newly formed diploid cell. The joining of an egg and a sperm or two spores.

We have discussed where the energy for life comes from but not what life is. We normally just accept it, but let us take a closer look. According to the dictionary: "Life is that property of plants and animals which makes it possible for them to take in food, get energy from it, grow, adapt themselves to their surroundings, and reproduce their kind." In order to do all that, a living organism requires various components to function together both mechanically and chemically. Most of these entities occur inside a unit called the cell. Therefore, we can consider a cell as the basic unit of most living things.

THE CELL

The word *cell* was first used in a biological sense 300 years ago by Robert Hooke when he used a new invention called the microscope to examine cork. Hooke was observing the cell walls, composed of **cellulose,** that surround most plant cells. It was not until 150 years ago that Purkinje, a Bohemian physiologist, separated all of the living material in the cell from the cell wall, which is a nonliving structure, and called it **protoplasm.** Today all of the many living components within any cell are included under the umbrella term of protoplasm. **Cell theory** includes the principle that the cell is both the functional and the structural unit of living organisms, both plant and animal. Included in the modern cell theory is the basic premise that a new cell can only come from the division of another cell.

There are two basic kinds of cells. One is primitive and believed by science to be similar to the first form of life on the planet. The kingdom Monera, which contains the true bacteria and the blue-green algae, has a structure of this type. This simple cell is called a **prokaryotic** cell. The second type of cell is found in most all other life forms and is more complex. It is called a **eukaryotic** cell. The main difference is that the eukaryotic cell has a much more complex structure, with membrane bound organelles, and more parts than do prokaryotic cells. (See Figures 5–1, 5–2.) This more complex cell is the type found in most living things with the exception mentioned above—the Monera.

Some of the basic parts of a primitive prokaryotic cell are the **cell wall,**

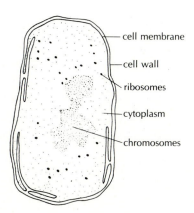

cell membrane

cell wall

ribosomes

cytoplasm

chromosomes

Figure 5–1   Prokaryotic cell.

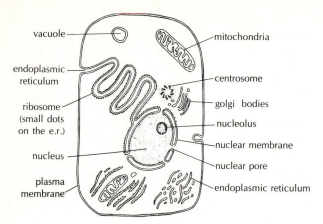

vacuole —
mitochondria
endoplasmic reticulum
centrosome
ribosome (small dots on the e.r.)
golgi bodies
nucleolus
nuclear membrane
nucleus
nuclear pore
plasma membrane
endoplasmic reticulum

Figure 5-2 An example of a Eukaryotic cell of an animal.

cell membrane, chromosomes, ribosomes, and **cytoplasm.** In the more specialized eukaryotic cell we find all of the above and much more. Most noticeable is the **nucleus** along with other constituent parts such as **mitochondria, vacuoles, golgi bodies,** and **endoplasmic reticulum.** In the nucleus are the chromosomes that contain the **DNA** (deoxyribonucleic acid), and the nucleolus which contains the **RNA** (ribonucleic acid). The DNA and RNA are the "program" for the next generation as well as for the present one. Any change that occurs in the DNA of a cell will be passed on to all generations that follow. One common way the DNA is changed is by having a naturally occurring x-ray from our sun pass through the chromosomes of a cell. When that cell reproduces, all of its progeny will carry the new information or "trait."

The cell wall is a poriferous nonliving case or shell around most plant cells. It is secreted by the cell and is composed of cellulose, which gives the plant its structure. It could be considered to be an external skeleton for each cell.

The cell plasma membrane is a living part of the cell and acts as a container for all the protoplasm. It is semipermeable and controls everything that enters or leaves the cell. Different factors control the passage of a substance through the plasma membrane. These include physical size of the molecule, electrical charge, number of water molecules, and even what the substance is soluble in. **Lipid** soluble substances seem to pass through easier. This is probably because the membrane has a layer of lipids as part of its makeup. Most plasma membranes have a three-layer structure of which the lipid layer is a part.

The nucleus is separated from the cytoplasm in the cell by a nuclear membrane. It can be located in different locations within the cell depending on the type of cell. Like the plasma membrane, the nuclear membrane controls the material in and out of the nucleus. It is a double-layer membrane and has the capability of selectively passing large molecules such as RNA through it. The nucleus is required for growth and reproduction of the cell. In the nucleus are located the chromosomes. These are composed of both DNA and RNA and are made up of units called **genes.** Every organism has a predetermined number of chromosomes in its cells. All the cells that comprise the organism, no matter what their shape or function, have the same number of chromosomes. For example, humans have 46 chromosomes. These occur as 23 pairs of similar chromosomes. If a cell has complete pairs of chromosomes it is said to be **diploid.** If a cell has only one of each pair of its chromosomes it is **haploid.** The body cells are diploid whereas the egg and sperm cells are haploid. When an egg cell and a sperm cell join, the cell becomes a complete diploid cell.

The nucleus of the cell also contains a **nucleolus.** Some cells have more than one nucleolus, but the number is constant within any given species. The main function of the nucleolus is to aid in reproduction of the cell and store the RNA. Without it the cell will not divide.

Other **organelles** found in the cytoplasm include the **centrioles.** During cell division, filaments form an **aster,** which radiates from the centriole, to join with the aster from the other centriole to form a **spindle** between them. The chromosomes line up on the spindle during cell division. Centrioles are found in animal cells and a few sperm cells of some plants.

The **mitochondria,** which vary in number and size, function to produce enzymes in the cell to convert food to useful energy for the organism.

The **endoplasmic reticulum** consist of small tubes and channels which run through the cytoplasm of the cell. The cell cytoplasm is divided into many compartments by the membranes of the endoplasmic reticulum, which allows separate functions to take place in the same cell. These channels also let substances pass through the cytoplasm.

Ribosomes are found in large numbers, sometimes in the tens of thousands per cell. They act in protein synthesis. All cells have them. They also contain RNA, which acts as a messenger to the rest of the cell from the DNA.

The **golgi bodies** act as storage areas for material manufactured in the cell. When enough material has accumulated it is released from the cell for use elsewhere in the organism.

A **vacuole** is a bubble in the cytoplasm that has several uses. Food can be stored and then broken down in the vacuole so it can be absorbed through the vacuole membrane into the cytoplasm. Waste material produced by the cell functions can be placed in a vacuole later to be released from the cell through the plasma membrane. The various uses of the vacuole are very important to the cell.

CELL DIVISION

Because of their structure and complicated physiology, cells are limited as to how big they can be and still function efficiently. In order to grow, an organism must increase the number of cells in its body. The process of cell division for the purpose of growth is called **mitosis.** (See Figure 5–3). This type of cell division is also responsible for repair of any cell that is injured or any cell that wears out. It is common in all living things except Monera. When a cell divides using mitosis the two "daughter" cells produced by the division are the same as the "parent" cell in every way except size. Mitosis, then, is the division of the nucleus and the organelles within it into two identical nuclei. This process works in concert with **cytokinesis,** which is the division of the rest of the cell. When division starts it is a smooth and steady process, but there are four obvious changes that can be easily observed. Scientists call them *stages.* They are **prophase, metaphase, anaphase,** and **telephase** stages. What follows is a summary of some of the activity in the cell that relates directly to mitosis during the four stages.

Prophase: During this stage the chromosomes become easily visible. They shorten, thicken, and divide lengthwise to become a pair. Each half of the chromosome is called a **chromatid;** two chromatids make a pair. The centrioles have migrated to opposite sides of the nucleus and are producing fibers called *asters.* Longer fibers, the spindle fibers, radiate to meet each other in the middle of the nucleus where the chromosomes have all lined up. The spindle fibers have attached to the chromosomes and the cell is now in *metaphase.* This phase or stage lasts only approximately one-tenth as long as prophase. It takes about three to five minutes. During metaphase the two chromatids separate and start to move apart along the spindle fibers.

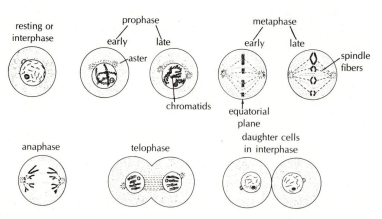

Figure 5–3 Mitosis is the process that somatic cells use to divide and grow.

This state is called *anaphase*. During anaphase one of the chromatids of each chromosome is pulled to one of the centrioles to form two daughter nuclei. This too is a fairly short phase, about twice as long as metaphase. The fourth and final division stage is *telephase*. During this stage a new nuclear membrane is formed around each daughter nucleus; the centriole splits so there are again two of them; the spindle fibers disappear; and the chromosomes become hard to see. At this point cytokinesis starts and the plasma membrane indents and pinches off into two new daughter cells, each with one of the daughter nuclei. This takes about as long as prophase. The cell is now in normal **metabolic stage** and can remain this way from days to years. This type of cell reproduction is responsible for such abnormalities as cancer if the cells start dividing faster than they should.

Tissues are groups of cells in multicellular organisms (**metazoans**), which have similar structure and function. An example is *epithelial* tissues. These cells cover the organism. Skin is an example of epithelial tissue.

Organs are groups of tissues that perform more complex functions, and groups of organs form **organ systems** for the most complex functions. The more advanced or modern an organism is, the more likely it is to have complex organ systems. The main systems are the integumentary, skeletal, muscular, nervous, endocrine, alimentary, respiratory, excretary, circulatory, and reproductive systems. All of these systems may not be found in any given species. For example, worms do not have a skeletal system, and protozoa have no organ systems.

The scientist who studies cells is called a **cytologist,** whereas one who studies tissues is a **histologist.** The study of organ systems is **gross anatomy.**

There are two functionally different types of cell division. Mitosis, which we have discussed up to this point, occurs in **somatic** cells, which are normal body cells that comprise all the tissues other than the reproductive tissue. The function of the **reproductive tissue** is to produce the egg in the female, sperm in the male, and spores in some plants. The eggs, sperm, and spores are produced by the process of **meiosis.** These reproductive cells are called **germ** cells. The only place in the organism where this type of cell reproduction takes place is in the **gonads** of the organism, and the only cells produced are either eggs, sperm, or spores. All other cells that make up an organism, whether plant or animal, are created by mitosis. Meiosis for some reason seems to be a difficult concept for most students; perhaps this is because it appears at first glance to be similar to mitosis but in reality is quite different. Meiosis produces **haploid** cells as daughter cells. A haploid cell has only one half the number of chromosomes that the parent cell has. This reduction process takes two divisions of the cell, so meiosis is really a pair of cell divisions. The end result is that, instead of two cells which are like the parent, there are four cells which are not like the parents. They are in fact different from each other. Each of them has a different chromosome

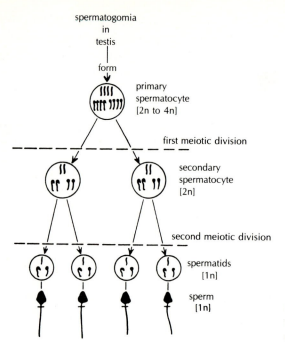

spermatogomia
in
testis

form

primary
spermatocyte
[2n to 4n]

first meiotic division

secondary
spermatocyte
[2n]

second meiotic division

spermatids
[1n]

sperm
[1n]

Figure 5–4   Spermatogenesis: the formation of sperm by the process of meiosis.

combination. That is why sexual reproduction produces young that are not exactly like the parent either in appearance or function. This difference is the basis for adaptation and evolution. No two sexually produced organisms react to their environment exactly alike. As the environment changes some members of any given species will adapt to the changes and some will not be able to. There are a number of new terms that must be learned in order to discuss meiosis.

Spermatogenesis and **oogenesis** are the process of forming (genesis) the sperm (seed) and the egg (OO). The sperm and egg are called sex cells or **gametes.** The term **gametogenesis** is used to mean formation of any gamete (sperm, egg, or spore). Somatic cells have pairs of chromosomes and are said to be *2N*, or diploid. Gametes have one chromosome of each pair and are *1N*, or haploid. When two gametes combine they form a diploid (2N) **zygote.**

Meiosis is different from mitosis in a number of ways but only the most obvious will be discussed here. During the prophase stage of the first division of meiosis, the chromosomes are duplicated and instead of a 2N condition the cell becomes 4N or **tetrad**—having four chromosomes. As the cell divides, each of the newly formed cells get a pair of chromatids during anaphase, instead of one chromatid as it did in mitosis. During the second meiotic division these new 2N daughter cells divide again to separate the two chromatids of the first division. The end product is now four haploid

(1N) cells. If the haploid cells are male (spermatogenesis), they normally have tails and can move through their environment. We call them *sperm*. If they are female (oogenesis), they are not able to move and we call them *eggs*. In plants there may be no visual difference and we call them *spores*. When an egg and sperm join, or when two spores join, we call the new 2N cell a zygote. The zygote then divides through mitosis to form first the **embryo** and then the mature organism. The mature organism has **gonads,** which produce gametes through meiosis. Thus the process repeats itself again and again.

REVIEW QUESTIONS

1. What are the differences between a prokaryotic cell and a eukaryotic cell?
2. What occurs during each of the four phases of a mitotic division?
3. Where does meiosis occur and what does it produce?
4. Why is cellulose important in a plant?
5. What parts of the cell are directly connected with heredity?

Part Two

ENVIRONMENTS
OF THE OCEANS

Figure 6–1 This crack in the rock is the home of these turban snails (*Tegula spp*). Because they are gathered in a small space and not found in a dense population in other areas nearby, we can call this crack a microenvironment. There must be some environmental facts or factors which are beneficial to them in this crack which they do not find only a few inches away.

Chapter Six

ENVIRONMENTAL

SUBDIVISIONS

DEFINITION OF TERMS USED IN CHAPTER 6

Alongshore current: A current, parallel to the shore, transporting the water accumulated by the incoming waves as they break on the shore.

Anaerobic: Living in the absence of free oxygen, which most organisms need for respiration.

Benthic life: Those forms that live on the bottom (epifauna) or in the bottom (infauna).

Nekton: Swimming pelagic organisms.

Neritic region: The area (water) over the continental shelf.

Oceanic region: The water beyond the continental shelf.

Pelagic: Any organism that lives in the open sea with no lasting contact with the bottom.

Plankton: Drifting organisms.

Substrate: The type of material the organism has available to settle on, such as rock, sand, or mud.

Zones

Euphotic zone: That part of the ocean where light is present in sufficient quantities to permit photosynthesis.

Dysphotic zone: That area beneath the euphotic zone where light can still be detected, but does not possess enough energy for photosynthetic activity.

Aphotic zone: Area beneath the dysphotic zone where there is no measurable light from the sun.

Epipelagic zone: The top 200 meters of the sea.

Mesopelagic zone: The zone from 200 meters to 1,000 meters deep.

Bathypelagic zone: The area from 1,000 meters to 4,000 meters deep.

Abyssopelagic zone: The area from 4,000 meters to 6,000 meters deep.

Hadopelagic zone: The area below 6,000 meters deep.

THE COASTAL ENVIRONMENT

Pragmatically, the coastal zone is the most important area of interest for the average person. This is the area that we fish from, sunbathe in, swim and surf in, scuba dive in, and visit just to relax. The more we understand it, the more we can enjoy it. The general topics we will discuss are its sand dunes, estuaries, and marshes, and zonation caused by wave impact, substrata and tides.

Sand Dunes

Sand dunes are not common in all areas. Dunes are formed in areas where the beach is flat for a reasonable distance behind the high tide level and where a good supply of sand is being transported along the coast by the along shore currents. The sand is worked up high on the beach by the spring tides. Litter, such as pieces of driftwood, is also pushed up. This litter acts as a shelter for some of the sand around it; and as the wind moves the sand, it piles up and soon covers the debris on the beach. A few species of plants will grow well in the loose sand. They will start to gather sand around themselves as they grow and produce a place where the sand is sheltered from the wind. As more plants grow, more sand is deposited and the sand level rises. Some dunes reach heights of over 16 meters (50 feet), although most are less than that. In some areas, dunes are seeded by adding straw and seeds of dune plants so a dune will build up and protect the area behind it. This was done at Chamber in Sussex, England, to stop erosion along the coast. An established dune is a very delicate ecology and can be damaged by people using it as a pathway. People walking over the dune break the root system of the dune plants and allow the sand to be blown away by the wind. Some dune areas have been placed off-limits to people in order to protect them. One such dune system is near Coal Oil Point in Santa Barbara, California. The area is used for research and study.

BARRIER ISLANDS

Any island that lies off the coast and acts as a barrier to the waves reaching the coast is a *barrier island*. These islands can be composed of sand, coral, or rock. If they protect the coast and create calmer water between themselves and the mainland they are barrier islands.

An example of sand islands occurs off the coast of Mississippi and extends into Florida on the coast of the Gulf of Mexico. The U.S. government has made this area a national seashore, which is a type of national park. Ten islands form a chain to protect the coast from Pensacola, Florida, north to Biloxi, Mississippi. Oddly, the entire future of these sand islands depends on a plant called "sea oats." So important are these plants that it is a crime to pick them. The sea oat has an elaborate stem and root system that holds the sand in place. If the sea oats were to die, the sand islands would be washed away by the storms. As it is, the islands change shape during a storm, but they stay intact and protect the coastal cities a few miles away from severe damage during hurricanes.

The sound, the body of water that formed between the islands and the coast, acts as a quiet area for birds, fish, and people. Birds feed here, fish breed here, and people play here. Without these islands the coast would be little use to humans because of the damage that would be done to it on a regular basis by annual storms. Fish and other wildlife of the area, which include over 100 species, would be drastically changed and reduced. Creating a national seashore and monitoring the use of this area is one of the good things our government has done in its effort to conserve our wildlife resources.

Estuaries and Marshes

Natural bays, estuaries, and marshes are becoming a thing of the past. Today we build boat harbors out of them by dredging or fill them in and build houses over them. Only recently has the public outcry brought laws that help protect our coastlines, at least in some areas of the country. Now we must take steps to keep pollution from destroying the ones we have left.

Although plants are the main life form of the marsh, these shallow, quiet water areas act as breeding grounds for many types of marine animals and grazing areas for land animals.

Where fresh water (which is less dense than salt water) runs into the estuary, a layer of low salinity water is often found near or on the surface. Also, silt carried by the fresh water runoff settles out to form mud flats. Because most runoff occurs during winter, the condition of lowest salinity is normally combined with the coldest temperatures, while warm water temperatures are combined with high salinity due to little runoff and a high evaporation rate of these shallow areas during summer. The less tidal flow in the bay, the more extreme these conditions are. Permanent residents then must have a wide tolerance for many factors, but particularly for temperature and salinity. Plants adapt better to these types of conditions. In 1956 Paviour-Smith found that only 2 percent of the total life of a New Zealand marsh was of animal origin.

ZONATION

The zoning of any area into smaller, recognizable areas is a way of making things easier to understand by delineating small segments to be studied at one time. A zone is any area with a given set of characteristics that sets it apart from other areas. Different researchers have used different characteris-

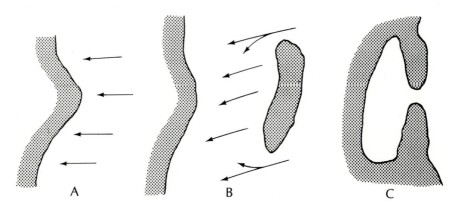

A. Wave action directly on the coast with no protection to break the force. This is called an *open coast*.

B. Wave action is disrupted by an island or any other obstacle. This is called a *protected outer coast*.

C. Total protection from wave action. This is called a bay. If fresh water runs into the bay from a stream or river, it is called an *estuary*.

Figure 6–2 Wave action and coastal formation.

tics from time to time; there are no distinct zones upon which everyone agrees. Although this indefiniteness may be confusing to the student who reads several books and finds different terms or areas defined, it is important to remember that zones are for our convenience and have nothing to do with the environment other than to make it easier to study. Consequently, we can divide the environment into any number of parts we wish and call them anything we like. Unfortunately, many researchers have considered it necessary to add many new terms to an already complicated scientific vocabulary. Often these new terms have apparently no greater significance than to satisfy the ego of their originator. The zones used here are as uncomplicated as they can be and still be consistent with the majority of books in use today. We can work with three obvious factors: **wave impact, substrate, and tides.**

Wave Impact

Wave impact breaks down into three easily recognizable subdivisions: heavy wave impact, moderate to low wave impact, and no wave impact. These subdivisions are commonly called **open coast, protected coast,** and **bay,** respectively. The influence on the life forms in each area is obvious. Open coast life forms must be smooth to cut down water drag on their bodies, must hold on tightly or burrow deeply to maintain position, and must be strong structurally to survive crushing or grinding action on their bodies. On the other hand, bay animals can be ornate in design and delicate in nature. The protected coast animals need some protection against the waves, but are rarely exposed to severe impact.

Substrate

The type of bottom will also influence what forms may survive on or in it. The most common substrates are mud, sand, rock and piling.

    Mud.   Mud is created by runoff from land that settles out in quiet water areas. It occurs almost exclusively in bays and estuaries along the coast, except for the mouths of great rivers such as the Mississippi where a muddy bottom extends some distance out to sea.

    The mud environment creates special problems for the animals that live there, one being a result of the small particle size of the material which composes mud. The small particles of mud and silt can penetrate the respiratory systems of many animals, and they suffocate. Only animals with specialized gill systems survive well in the mud environment.

    Oxygen levels normally are low in the sediments because the fine particles are so closely packed that virtually no circulation takes place beneath their surface, and the organic matter settling there decays and uses up available oxygen. Much of the life activity under these conditions is carried out by bacteria. Beneath the surface of the sediment, much activity is without oxygen, a condition called **anaerobic.**

Sand.   Sand composition varies with the environment. On Hawaii much of the sand is pulverized lava. In the Caribbean on Cozumel, the sand is pulverized coral. In most parts of the United States and Canada, it is composed of quartz and feldspar. Sand areas allow the water to percolate or flow between the particles, so there is generally good oxygenation at some depth in the sand beach. Because the larger particle size also allows for quick drying, the upper levels of the beach are quite dry, and the lower levels dry rather quickly as the tide recedes.

Each wave moves sand as it comes ashore. This movement of sand particles tends to act as a giant grinder. The animals that live in this area must either have heavy shells, move with the sand, or burrow deep into it, to avoid being ground up.

Rock.   As a place to live along the shore, the rocky areas support many types of organisms. When examining them, one is overwhelmed with the large numbers of different types of life found there. The main reason for this great proliferation of species is the variety of microenvironments found on a rocky shore: cracks, crevices, exposed areas, tide pools, etc. Rocky areas have good oxygenation, good food supply and, generally, areas of good protection. The species that live here usually attach themselves to surfaces. Although a few bore into the rock, most hang on with a sucking foot, like the snail; hide in the holdfast of the algae, like the worms as they cling

Figure 6-3   A colony of mussel growing on one rock with very few on the surrounding rocks is one of the ecological phenomena that is interesting to explain.

Figure 6-4 As the tide comes in, the rocky shore is flooded and all the organisms have new food and water brought to them. The organisms' periods of exposure to air between the high and low tides depend on their location on the shore.

to the crevices; or "glue" themselves to a rock with a type of cement, like the barnacles.

The rocky shore is the place to which people are drawn at low tide to peer and probe at their local marine life. In heavily populated areas, the "peer and probe" group have all but destroyed the majority of their tide pool areas. Along the California coast it is now illegal in most areas to take or disturb any tide pool life. Although the laws have helped, the many crushing feet of just the "peerers" do almost as much damage as "probers" used to do.

Pilings. The wood pilings of piers, docks and other artificial structures are noted as a separate environment because they support some types not found elsewhere. The boring clam called the **Teredo** is the best example. It is a wood-boring bivalve of the phylum Mollusca, and not really a worm at all although it is often called one. It is responsible for much damage to pilings, ship bottoms constructed of wood, and logs, rafted, ready to go to market. Normally pilings are vertical in the water, which allows for easy determination of the tidal zones discussed later in this chapter. A piling, therefore, makes a good study area for the basic student because of this sharp delineation of recognizable zones.

Tidal Zonation

The rise and fall of the tides periodically expose the organism along the land-sea interface to air. The regular exposure pattern creates a very sharply delineated zonation in most areas, which is most easily recognized on pilings and rocky shores because of the more or less fixed species assemblages that live there. The distribution of species in this area of zonation is dependent upon their ability to withstand being left high and dry when the tide goes out. In most cases we can identify four zones based on the average range of high and low tides and the extreme range of high and low tides.

   Zone I, the highest and driest, is sometimes called a spray zone or upper littoral zone. This zone is dry most of the time. It is above the average high tide and is wetted only by spray or during high spring tides. The periwinkle (*Littorina*) is a common inhabitant of this area. The species found high in the rocks have adapted so well to exposure to air they develop difficulties in respiration when left under water for long periods of time. The little barnacles (*Balanus*) and (*Chthamalus*) are also common in this zone.

   Zone II is exposed normally twice a day during low tide. It falls between the top of the average low tide level and the top of the average high tide level. Its exposure to air lasts 4 to 6 hours at a time. The black turban snail (*Tegula*) is common along with several species of acorn barnacle and limpets, depending on the area. This is the most common "tide pool" area known to the average rocky shore enthusiast.

   Zone III is covered the majority of the time. It is located between the top of the average low tide and the *bottom* of the average low tide. Although it is exposed during many low tides, the period of exposure is short, generally 1 to 3 hours. Acorn barnacles, gooseneck barnacles, some common sea stars, and many other invertebrates are found here. More types are found as less exposure to air is encountered. Zones II and III are also referred to as the mid-littoral zone and the barnacle zone.

   Zone IV falls below the average low tide and is exposed to air only during the very low spring tides. Its exposure to air is short, only an hour or two, and occurs at irregular intervals. Generally this zone is exposed during the same period of the month that Zone I is wetted due to extreme tides. Many organisms occur here, including many of the marine plants. Some

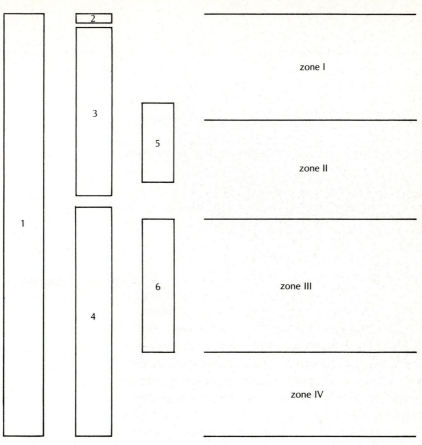

1 = tidal range
2 = spray zone
3 = extreme high water range (spring tides)
4 = extreme low water range (spring tides)
5 = mean high water range
6 = mean low water range

Figure 6–5 Zonation holds true in relationship to the tides regardless of the tidal range or location geographically.

refer to this zone as the sublittoral or lower littoral zone, and in sandy areas it has been called the submerged beach. The best way to recognize this zone in an area is not by what is in it, but by the generally sharp line above which its organisms do not occur.

By combining the information above we can describe various environments so that our audience will understand a great deal about the organisms that live there. For example, we can describe an environment as being in Zone III of an open coast on a rocky substrate. Although brief, this description tells us much about the physical appearance, the physiology, and the general life style of an organism found there. Complete understanding of

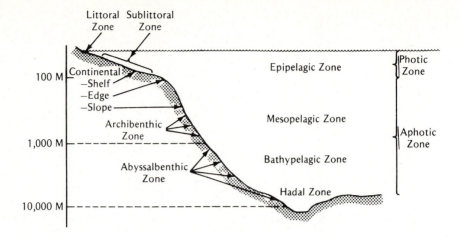

The division of the marine environment is commonly made in the terms and areas given above.

Figure 6-6   Some divisions of the marine environment.

this method of description is a requirement of the student, as it will allow a sensible pattern to develop in the mind of the student studying a local environment or reading about others.

OFFSHORE DIVISIONS

The subdivisions we have been discussing all occur in the first few meters along the edge of the ocean. The rest of the ocean must also be divided so we can use a similar short description when referring to a segment of it as

Figure 6-7   A beach hopper, one of the upper zone arthropods, is a common resident of many sand environments.

Figure 6–8 Because the upper zone is wetted on very high tides only, organisms that live here are well adapted to exposure to air. Here we see kelp that has washed ashore and been left behind by a high tide.

an environment. Different factors are used to identify particular portions in the open ocean because the factors that are convenient along shore have little or no importance at sea. The most commonly used physical factors are depth and light intensity. Animals and plants that live in the open sea and have little or no contact with the bottom are called *pelagic* species. They swim or drift their entire life. The term used to define the drifters is *plankton;* the swimmers, *nekton.*

Neritic and Oceanic Regions

The waters that lie over the continental shelf are in the neritic region. As discussed earlier, the width of the continental shelf varies in different parts of the world, so the neritic region varies to the same degree as the continental shelf. The water that lies beyond the edge of the continental shelf is called the *oceanic region.* By definition, then, the neritic waters are no deeper than 200 meters at the edge of the continental shelf and become shallower as they get closer to shore. The oceanic waters start over the outside edge of the continental shelf, so they are much deeper, extending to the deepest part of the ocean floor.

Light Intensity

The water in the ocean acts as a filter to the sunlight that shines upon it. The degree of filtration depends on the amount of particulate matter suspended in the water. The more silt or plankton in the water, the more difficult it is for the light to penetrate through to a deeper depth. In some water near the mouth of rivers, particularly after a storm, light may penetrate less than one meter. In the clear water of the open sea, it is possible that light can be detected as deep as 600 to 800 meters. Because the majority of the ocean is away from shore, the general zones are based on clear, open water. Although light may penetrate to these greater depths it is not intense enough for plants to photosynthesize. This gives us a good indicator to define a zone. The surface water where light is intense enough for the planktonic plants to produce food through the process of photosynthesis we call the *euphotic zone*. The area beneath the euphotic zone where there is light but it is not strong enough to produce a photosynthetic reaction, we call the *dysphotic zone*. Still deeper water where no light penetrates we call the *aphotic zone*. Because the light intensity varies in different parts of the ocean, we have based a system of zonation on depth so we could be as consistent as possible. These depth zones take into consideration the photic zones as well as other factors.

Depth Zones

The *epipelagic zone* extends from the surface to 200 meters. This corresponds roughly to the euphotic zone and thus it supports most of the life in the sea because of photosynthesis. The *mesopelagic zone* is from 200 to 1,000 meters. This corresponds roughly with the dysphotic zone.

The *bathypelagic zone* extends from 1,000 to 4,000 meters. The *abyssopelagic zone* extends from 4,000 meters to 6,000 meters, and below 6,000 meters is called the *hadopelagic zone*. The bathypelagic, abyssopelagic, and the hadopelagic zones all fall within the aphotic zone.

Bottom Zones

In contrast to the pelagic species, the species that live on or near the bottom are called *benthic* forms. We divided the benthic organisms into two groups according to where on the bottom they live. If they live mostly *on* the bottom, as do some fish or sponges, they are called **epifauna.** If they live *in* the substrate as do some clams and worms, they are called **infauna.** We use the same depth zonation for the benthic zones as we do for the pelagic zones. Any bottom animal found at depths between 1,000 and 3,000 meters would be a bathybenthic organism; if it were found below 6,000 meters, it would be a hadobenthic organism. Like the terms used for inshore zonation, the terms used for oceanic zonation tell us a great deal about a specimen. Describing a given specimen as an example of the infauna of the abyssobenthic zone would tell us the depth and general life style of the specimen.

REVIEW QUESTIONS

1. Why is it necessary to divide the ocean into zones?
2. What effect does wave impact have on a coastline?
3. Why is sand a difficult environment for most animals and plants?
4. What is the main criteria for separating the tidal area into tidal zones?
5. Why is light intensity in the water important enough to be used to create zones?
6. Why are "sea oats" important?

Chapter Seven

THE DRIFTERS

DEFINITION OF TERMS USED IN CHAPTER 7

Bio-mass: The total amount of biological material found in the environment.

Bloom: An abnormally dense population of phytoplankton caused by rapid reproduction under optimum conditions.

Holoplankton: Organisms that exist only as planktonic forms.

Macroplankton: The larger members of the plankton, such as jellyfish.

Meroplankton: Those planktonic organisms which are planktonic forms only during their larval stages.

Metamorphosis: Change in form, structure or function—physical transformation more or less sudden-development after the embryonic state—as from tadpole to frog.

Microplankton: Plankton that can be collected in standard plankton nets.

Nanoplankton: Larger than ultraplankton, but too small to be caught in nets of the finest mesh.

Phytoplankton: Those organisms within the plankton that are plants.
Ultraplankton: Bacteria-sized plankton.
Zooplankton: Those organisms within the plankton that are animals.

The waters of the world are filled with small drifting organisms. The majority of all the major and minor groups of animals and plants that live in a water environment is represented in the plankton. The scientist groups these drifters together and calls them **plankton.**

GROUPS OF DRIFTERS

Marine plankton is so varied in its makeup that we must separate it into several major groups for easier study. One grouping is based on whether the organism is plant or animal. The plant drifters, called **phytoplankton,** are mostly very small one-celled types. The animal drifters, called **zooplankton,** range from microscopic to over 1,000 pounds. Plankton, then, does not have to be microscopic although most of it is.

Another method of grouping is by size. The smallest of all plankters are called *ultraplankton,* which consist of forms such as bacteria. These organisms are so small that collecting them is extremely difficult. They are studied by microbiologists.

The next size grouping is called **nanoplankton.** Nanoplankton is so small that it will pass through the mesh of our finest-meshed nets. This makes their study difficult for the student biologist. To collect nanoplankton, the researcher must pass the water through a fine filter or spin it in a centrifuge.

The plankton that can be filtered out by use of the fine-meshed plankton nets is called *microplankton.* These microplankters range from the small one-celled plants, which must be observed with a microscope, to small ani-

Figure 7-1 Some planktonic forms look almost like spaceships as they drift along with the current.

Figure 7–2 Comb jellies found near the surface at night and early morning, drifting in the plankton.

mals quite easily seen by the unaided eye. This group is the main source of bio-mass production in the sea.

The largest plankton forms are called the **macroplankton.** This group includes the various jellyfish, ocean sunfish (*Mola mola*), and other larger forms.

One of the more interesting facts about marine plankton is that at any given time, there are a great many larval forms of larger animals in it. Barnacles, crabs, fish, and many others have larval forms that live for a short time as members of the planktonic community. The majority of these planktonic larvae bears no resemblance to the adults. Some have hairlike appendages; others have paddles to move themselves through the water. As the benthic types mature, they sink to the bottom and assume their adult form and life style. A large proportion of these short-time planktons undergo a complete change from egg to adult, which may take them through several stages of entirely different body shapes. This type of development, in which an organism goes through different body forms before the adult stage is reached, is called *metamorphosis*. These early life plankton forms are called the **meroplankton.**

Organisms that exist only as plankton are also greatly diversified in the ocean. This group, called **holoplankton,** includes members from most of the major divisions of animal life as well as many plant forms. The salt water of the ocean is denser than fresh water. Because this denser water will support more weight, heavier objects that would sink in fresh water will tend to be neutral. Consequently, in salt water more forms float and exist as plankton rather than sinking to the bottom. In salt water there are mollusks, chordates, and some worms that have adapted to life in the planktonic communities as members of the holoplankton.

It should be clear by now that the planktonic community is very large, highly diversified, and divided by the scientist into many groups. Study of the small plankton is, biologically speaking, relatively new because of its size and our technology for studying it. Only in recent times have we understood the importance of its role in primary production and as a breeding ground.

It is suspected that the new chemosynthetic organisms of the deep-sea vent habitats have colonized these hydrothermal vents by using the planktonic mode of travel from vent to vent. Much more research will be needed to verify if this is true or not.

All of the above groups are drifters, collectively referred to as plankton. The organisms that do not drift but swim to control their placement in the sea must also be grouped for identification. The swimmers, called **nekton,** will be dealt with in a later chapter where we discuss their major groups. Most of the nekton fall into the vertebrate group, with representatives from others, such as the mollusks, to a lesser extent.

CONDITIONS AFFECTING POPULATION DENSITY

Because the plankters, by definition, are drifters, generally they are moved along by currents from one place to another, but any large bay or semiclosed area, such as the Gulf of California between Baja California and mainland Mexico, will have a standing population that shifts with the tides but may never leave the area. These standing populations are fed by nutrients washed off land or brought to them by the upwelling of deeper water. Because the conditions in such a defined area are more predictable, the plankton of the area is also more predictable. In the open sea where environmental conditions change, it is more difficult to predict which organisms will make up the planktonic community at any given time, place, or depth. We have said earlier that the densest populations of plankton are generally in the first 200 meters. This is true if we count all plankton, but not if we are selective as to type. The densest populations of small copepod crustaceans, the most significant zooplankton, are normally found at a depth greater than 200 meters. It stands to reason that the deeper the organism lives, the less food is available to it because the sinking detritus decreases as each layer of zooplankton in turn takes its share. These deeper plankton are not affected by light, but rather by density of the water and pressure. The various photosynthetic types that use light to produce food all have an optimum light intensity in which to function. This optimum light intensity is different for each species. Light that is too bright will drive them deeper; light that is too dim will bring them closer to the surface. Thus during the day when light is bright, the various species will layer themselves at different depths according to their optimum light intensity.

When the phytoplankton encounter an area especially rich in nutrients, it may multiply more rapidly than normal and produce a very heavy **bloom,** or extremely dense population. When this happens, the zooplankton and

Figure 7-3 As the plankton drift through the water, millions of predators, like these colonial cnidaria, catch and eat them as fast as they can.

even some fish will normally avoid swimming up into the dense phytoplankton. There seems to be a toxic material secreted by these heavy blooms that acts as a deterrent to the other species. One such heavy bloom is called *red tide*. Although the organisms that make up the red tide can vary, they all contain a red pigment, which in high concentrations actually colors the water.

Toxins produced by some of the organisms associated with the red tide, if concentrated, are poisonous to humans. Some shelled animals, such as sea mussels, filter-feed on the red tide organisms and concentrate the toxin in their own tissue without causing damage to themselves. If people then eat the mussels, the toxins will make them sick or even kill them. Consequently, there is a quarantine on certain shellfish during the time of the year that the red tide is generally found in the area. The red tide off the coast of Southern California in years past generally bloomed for about two months a year during the summer. In recent years, it has sometimes lasted five or six months. Many biologists in the region believe that the increase in organic

material brought to the area by a larger human population, and the con-
sequent increased sewage, account for the increase in the bloom's duration.

We have known for years that Antarctic waters are rich in plankton.
Although many factors, of course, figure in this abundance, we will take a
simplified look at only one, phosphates. Phosphates, one of the essential
factors in phytoplankton production, is the fertilizer of the oceanic garden.
Because of the geographic and hydrographic conditions in the antarctic region, the surface currents normally move northward, and deep water rises
to replace the surface water that is moving away. Rich in phosphates, this
deep water carries it up into the lighted waters where the phytoplankton
can use it for photosynthesis. The concentration of phosphates is approximately 50 to 60 milligrams per cubic meter of water. As the water moves
north to what is called the *Antarctic convergence,* which is at approximately
50° south latitude, it remains rich in phosphates. At the convergence, this
surface water starts to sink under the warmer, less dense waters of the south
Atlantic. North of the Antarctic convergence, the phosphate level drops to
25 to 40 milligrams per cubic meter. The lower levels occur in the summer
when the water of the Atlantic is warmer and less mixing can occur at the
convergence because of the greater difference in density. This level of 25 to
40 milligrams of phosphate per cubic meter is still rich, excellent for phytoplankton development. As we go farther north to around 40° south latitude,
we find another convergence area called the *subtropical convergence.* In
the subtropical water, the phosphate level drops to near zero; still farther

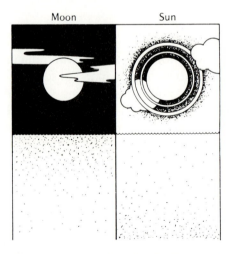

Planktonic organisms have optimum light levels. During the day they go deep
to reduce the light intensity. During the night they move to the surface to get
all the light they can.

Figure 7–4   Movement of plankton towards optimum light levels.

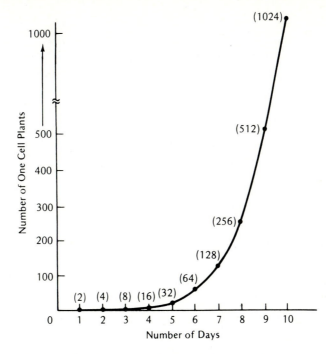

Many phytoplankton divide once each day, if this growth was not checked by predators and other environmental factors, they would fill the ocean in a short period of time.

Figure 7–5   Plankton growth.

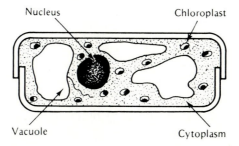

The basic shell design of most diatoms allows for expansion and contraction as the shell halves slip together or apart. They contain a nucleus, chloroplast, cytoplasm, and one or more vacuoles.

Figure 7–6   The basic diatom: primary producers of the sea.

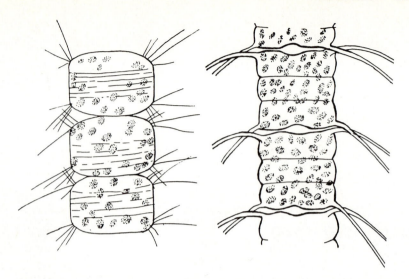

Figure 7-7 Some common diatom shapes. Often these diatoms are linked together in long chains.

north in the tropical waters, there is no phosphate in the water at all. As soon as it occurs from waste materials of other plankton and animals, it is consumed again. The density of plankton has been shown to have a high level of correlation with the availability of phosphate. The more phosphate, the more phytoplankton. Obviously, any of the factors needed for growth and reproduction can be the limiting factor to control the population. The decreasing amount of phosphates as we approach the tropics from either the north or south is a main factor in the paucity of the planktonic community in tropical waters, which in turn is a main factor in the traditionally clear surface water of the tropical regions of the world. This small example illustrates the great interrelated network of factors that function in the oceanic system.

REVIEW QUESTIONS

1. Why is size used to separate groups of plankton, rather than types of organisms?
2. What are the pros and cons of an organism having its larva form a part of the plankton for a portion of its developmental stages?
3. How does upwelling help the plankton?
4. Which group of zooplankton is the most significant, and why?
5. What is "red tide"?

Chapter Eight

SOME BENTHIC

ENVIRONMENTAL

CONSIDERATIONS

DEFINITION OF TERMS USED IN CHAPTER 8

Arenicola: Common beach annelid.

Astrorhiza: Territorial benthic foraminifera.

Balanus: Acorn barnacle, indicator organism for mid-tide zone.

Chthamalus: Acorn barnacle, indicator organism for mid-tide zone.

Donax: Bean clam.

Emerita: Mole crab.

Epiflora and Epifauna: Organisms that live on the bottom.

Euzonus: Common beach annelid.

Indicator organism: An organism whose life style is so well known that its presence in a new environment helps to indicate the ecological factors present.

Infauna: Organisms that burrow and live under the surface of the bottom, as opposed to epifauna.

Latimeria: Coelacanth, fish thought to have been extinct.

Littorina: Periwinkle, a Zone I indicator organism.

Macoma: Territorial benthic bivalve.
Myxophyceans: Common algae found at the zone I level.
Neopilina: Mollusk thought to have been extinct.
Panope: Geoduck clam.
Stylatula spp.: Sea pen.
Tresus: Gaper clam.
Tagelus spp.: Jackknife clam.
Tetraclita: Acorn barnacle, indicator organism for mid-tide zone.
Tivela: Pismo clam.
Zostera: Eel grass.

The benthic environment includes all organisms that live on or near the bottom, the **epiflora** and **epifauna,** or under its surface, the **infauna** of the sea. Two ways to divide this environment for study that are easy to define are by photic zone (with light) and aphotic zone (without light) and by hard substrate and soft substrate.

The photic zone of the benthic community all occurs on the continental shelf or on islands or sea mounts that rise within several hundred feet of the surface. The aphotic zone is generally deeper than 200 meters and includes the deep ocean bottom.

THE PHOTIC ZONE SOFT SUBSTRATE (BEACHES AND BAYS)

Sand beaches occur all over the world. Some of them stretch for miles without interruption, while others are reduced to small pockets of sand tucked away in the back of some rocky cave. In some areas, the sand is composed of **silica,** fragments of rocks from decaying land masses washed down to the sea by rivers and streams. Elsewhere, the beaches are composed of shells ground to small fragments by the surf as it pounded on the reef and on shore. There are also black sands where lava beds that flowed into the sea have broken down into fine particles. In the areas where rivers and streams flow from large land masses to the ocean, much silt is also carried to the sea. This silt creates mud layers in waters calm enough to allow the fine particles to settle to the bottom. These quiet waters are found beyond the influence of the surf. Fine particles may be carried into the sea by rivers, like the Amazon River, Mississippi River, or any other major river of the world.

Because of the difference in particle size between silt and sand and the water conditions that accompany either type of substrate, the life forms in each type differ considerably. The quiet water necessary for the silt to settle out and the soft texture given to the mud by the small size of its particles

Figure 8–1 This picture of the bottom of a tide pool shows several benthic organisms: the arthropod in the center; the Mollusca; the colony of bryozoans, which look like a honeycomb in the center of the picture; and several types of algae. They all exist on the bottom and, therefore, are referred to as benthic organisms.

creates a perfect environment for soft-bodied organisms with little or no protection other than the ability to dig a hole or burrow a tube to live in. This quiet environment is also a habitat where a shelled animal can develop a very ornate shell with long projections, without trouble from the force of water against an increased surface area. One major adaptation that is essential to live in such an environment is the ability to breathe without the gills becoming clogged by the small mud particles. In contrast, the environment of sand particles generally is associated with water movement of some sort. Wave action is the most common. The water movement and the force of the waves have a direct effect on the size of the individual sand particle. The stronger the force of the water (the bigger the waves), the larger the sand particle must be in order not to be carried away. When you find a beach with very coarse sand, it is generally a sign that big surf pounds on it at least occasionally. This water movement and larger particle size create a grinding action. The organisms that can tolerate this environment are generally very streamlined to cut down the surface area of their body exposed to the pressure of the moving water. They also must have some means of withstanding the grinding of the particles moving with the water. To do this, some have heavy shells; others orient themselves so the smallest area of their body is exposed to the particle movement; and others burrow beneath the shifting sand.

Most creatures do a little of all three in order to survive. In the sand environment, oxygen is rarely, if ever, a problem. The moving water mixes the oxygen-rich surface water and the particle size of the sand, as well as the movement of the sand by the water turbulence, allows a gas exchange at considerable depths under the sand. In the quiet waters of a mud environment, where water mixing and particle movement are minimal, the particle size is generally so small that circulation among the particles is impossible. Within a very short distance, only an inch or two, the decaying organic material uses up all of the oxygen and there is none available for use by animals. To live in this environment, some creatures have become anaerobic and create their oxygen through a chemical process. Others bring oxygen to themselves by keeping fresh water flowing through their burrow.

A Few Examples of Sand Dwellers

Many of the organisms familiar to us in the sand surf environment are "in-surf" species. That is, they live in the surfline and therefore are visible to the average beachcomber. Several of these animals travel up and down the beach with the tides so as to always maintain the same approximate position in relation to the surfline. One such animal, a representative of the phylum Mollusca, is a small bivalve of the genus **Donax.** This little clam, rarely over 3 centimeters long, has the ability to push itself up and out of the sand very quickly and likewise to dig into the sand very quickly. Its preferred location is at the upper edge of the surfline. To maintain this relative position, it has the ability to "feel" the incoming tide. It lives in areas of surf and uses the force of the waves as an indicator as to when it should rise to the surface and be carried up the beach to a higher position in the sand, where it digs in again. It is so sensitive that it will only pop out of the sand if the waves are big enough to indicate the incoming tide. It can be made to pop up by stamping on the sand, which stimulates the thud of the crushing breaker. It also pops up and is carried back down the beach as the tide goes out. This stimulus is obviously different, probably the degree of wetness or dryness of the sand in which it burrowed.

The phylum Arthropoda also has a representative that travels with the surf up and down the beach, the small mole crab (**Emerita**). The body of these crabs is generally under 3 centimeters long with a hard, smooth shell covering its back. This little crab burrows into the sand backwards so that its head and antennae are at the surface level of the sand and obtains food from the water rushing out from each surf wave. Because it uses the receding water to carry food, it tries to stay in a position on the beach where it has a maximum amount of water flowing over it at all times. This position requires that it move as the tide comes in and goes out. Able to dig in within one second, it allows itself to be washed up on the beach and then digs in. When the water recedes and the amount running over it lessens, it pops up, washes back down the beach a few feet, and digs in again. If conditions are right, thousands of *Emerita* are generally easily observed making their movement in the surfline. They are often caught by anglers, either by hand or with small wire baskets dragged in the surfline, to use as bait in surf fishing.

The genera **Arenicola** and **Euzonus** represent the annelid worms in the sand environment. Having soft bodies, these worms burrow under the sand and are part of the infauna of the beach area. Of the many species, some of them are beach dwellers and survive by digging a foot or so under the sand. They are detritus feeders and dig and feed at the same time by taking in sand, digesting the organic material, and passing out the remainder. Only 0.5 to 1 percent of the material taken in is organic matter; therefore, they filter around five times their body weight in sand a day.

On the beach in La Jolla, California, the density of *Euzonus* was 2,500 per square foot of sand surface. These creatures are easy to see when dug out of the sand, even though they are only one and one-half inches long, because of the hemoglobin in their blood that gives them a red coloring. The presence of a detritus feeder in the sand indicates there is an abundance of organic material to feed on that is not obvious to the casual observer. In just a quarter mile of beach, with a heavy colonization of these "blood worms," they will, in one year, eat and filter 3,500 tons of sand and find some 37 tons of organic material in it to digest. This comes to about 3 ounces of filtered sand per worm covering a population of nearly 40 million.

Even the flowering plants, which are so well developed on land, have a representative that can survive small and moderate surf areas. The genus **Zostera,** commonly called *eel grass,* is a common form in many places in

Figure 8–2   Scuba divers do a great deal of scientific survey work.

the world and is often exposed at extreme low tide. It grows in beds that cover an area from a few square yards to a few square miles. Many creatures use eel grass as their main habitat.

As we move into deeper water behind the surfline, many other forms are common. In California, the pismo clam (**Tivela**), is well known because of its food value. The gastropods of the genus **Olivella** are also found a little further out where they can burrow under the sand and escape the breakers. Representatives of the *cnidaria* are also found in this harsh area, which is rather surprising because most of this group have soft and unprotected bodies. The hydroid colonies found in the shallow sand areas either grow on the shells of other animals living there or, like the sea pansy, are anchored directly in the sand by a rootlike extension of their colony. Sea stars of several types and sea cucumbers are also common beyond the surfline. As in most environments, representatives of all the major groups can be found if the time is taken to look for them.

A Few Examples of Mud Dwellers

Where filter feeders are common in areas of good water movement, such as surf areas, deposit feeders are common in mud substrates. The fine particle size of the mud tends to clog the gills and filters of filter feeders. The mud dwellers are commonly found to ingest the mud and digest the organic material out of it. Worm-type animals are well suited for burrowing in mud and are very common in samples of mud bottoms. Again, all groups are represented, but it is unusual to find the same species here that are found in the sand environment. In general, the habitats require different life styles for survival. The sea pen (**Stylatula**) is an example of a bay form found in Newport Beach, California. It lives buried in the mud, anchored there by a bulbous posterior end. When the animal is covered with water, it extends above the surface of the mud and opens like a feather sticking up. When the water goes out during low tide or if it is disturbed, the entire animal can draw back under the mud.

Many clams use the mud as home, the quiet waters and soft substrate being ideal for their life style and anatomy. Clams such as the gaper (**Tresus**) and the geoduck (**Panope**) need quiet waters because their shells are not large enough to protect the entire body, and coarse abrasive sand would kill them. Thin-shelled types such as the jackknife clam (**Tagelus**) are very much at home in this unshifting substrate.

In any given substrate, one species generally dominates in number. This species is sometimes called an *indicator organism*. Once it is studied and its life style known, its presence as the dominant species in some new area being sampled tells the sampler a great deal about the conditions present.

Most of these benthic types have some area of territory that they maintain. Some territories are small, like that of the foraminiferan (**Astrorhiza**), which has a mucus net resembling a spider web about 6 millimeters in diameter; other territories are larger, such as that of the bivalve **Macoma,** which can extend its incurrent siphon some 15 times the length of its 7-centimeter shell to search out food. Thus it can extend its siphon to cover a meter; by circling around, it can feed off an area 2 meters in diameter—a very large feeding territory for such a small organism.

Zostera (eel grass) is a common plant found in mud as well as sand. It is not a marine alga, but rather a land-type seed plant. It has roots, whereas algae do not. These roots trap the particles of mud and create a very important progression in the step-by-step process in the succession of a bay to a dry land environment.

We would expect, as with almost all biological communities, that the number of different species within the community would increase as we leave the polar region and approach the equator. Although this is true with the epifauna, it does not seem to hold true to any significant extent with the infauna. In 1957 Thorson found that the number of species in the epifauna increased from 100 in the polar region to 350 in the temperate region to over 700 in the tropical region. While this increase in epifauna was taking place, the infauna remained the same, about 40 species in all locations. This

Figure 8–3 Highly desired for food, various species of abalone are found on rocks from the intertidal zone to depths of 150 feet. The often extensive growth on their shell, which makes them difficult to recognize, creates a space for some other organism to live in.

Figure 8-4 The benthic environment is home for a very large group of organisms of many different types.

constancy suggests that living conditions are more stable under the mud surface than on top of it.

PHOTIC ZONE-HARD SUBSTRATE (ROCKY SHORES AND CORAL REEFS)

When discussing the soft substrate we mentioned detritus feeders because detritus accumulates in the sediment for them to feed on. By contrast, in the rocky environment some type of water action generally keeps sand and mud from settling in one place for any length of time. There are rocky outcroppings in bays, of course, where water may be quiet, but they generally have a layer of sediment that has settled around the same area. Rocky environments usually occur in the littoral zone because the wave action, surge, and tidal currents do not allow sediments to settle. The rocky substrate of most interest to the average person is along the shoreline. Some widespread general features of rocky environmental zonation are readily recognizable. There is always a transition area from water to land, called Zone I in Chapter 6. The genus *Littorina* is perhaps the most widespread animal type of this zone. This small periwinkle snail has great resistance to drying out and can survive this transitional environment better than any other animal form. There are also a few plants that can tolerate this transitional area. These plants, black and of an encrusting nature, create an apparent black line along the rocky shore at extreme high tide level. They are generally of the group myxophyceans, or sometimes of the Verricaria-type lichens. This black stripe is characteristic worldwide to one degree or another.

Below Zone I is the normal intertidal area that is covered and uncovered by the normal neap tides. It is this area that has caused some confusion. Some observers have divided this neap tide area into as many as four separate zones; others consider it one zone. The place of observation is an important consideration. Off Southern California this area can be split into two zones (Zone II and Zone III) quite realistically, while in the Antarctic, only one zone is evident. In British Columbia, Canada, or San Felipe, Mexico, this area could be divided into three zones quite easily. Whether it is divided into one, two, or three zones, however, certain characteristic animals are found here. The acorn barnacles are the prime indicator organism. The three common genera are *Balanus, Chthamalus,* and *Tetraclita.* In fact, this is often called the barnacle zone.

Below the neap tide area is the spring tide area. (Zone IV) This low water zone has only occasional exposure to air on low spring tides for very short periods of time. This lower zone, being wet most of the time, has a much more variable population of plants and animals. The calcareous algae of the family Corallinaceae are widespread encrusting organisms of this zone. The temperatures in different parts of the world dictate the various other common forms of their zone. For instance, in colder water the form **Laminaria,** a brown algae, is common. In warm coral reef areas the rich growth in this zone is generally outstanding. Due to the zonation, the wave impact and all of the variable features of this in-between world of ocean and land, the diversity found here in environmental niches encourages a great number of different forms to make use of all available space. This is why the inshore portion of the sea has a greater number of species and variations of form than does any other area of the oceans.

DEEP APHOTIC BENTHIC CONDITIONS (MUDS, SILTS, AND OOZES)

There are rocky outcroppings and entire mountain ranges like the mid-Atlantic ridge, but the hard substrate is a small portion of the total aphotic benthic area. As discussed earlier, the deep ocean bottom is composed of **calcareous ooze** (mainly foraminifera), **siliceous ooze** (mainly diatoms), and **red clay** in the deeper areas where extreme pressure makes the solubility of the others high. It is estimated that these sediments accumulate at the rate of about one centimeter in a thousand years. Despite this very slow rate, the several hundred milllion years the ocean has existed has made these sediments quite deep in places. The animal life found here is directly related

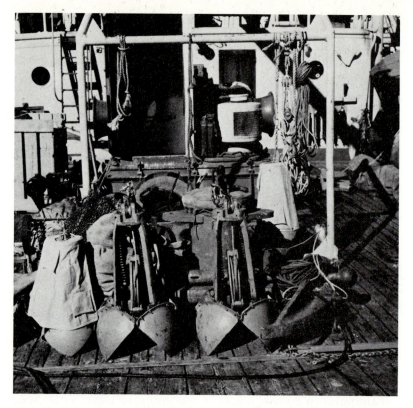

Figure 8-5 The orange peel grab is one of the many sampling devices used to bring up a piece of the ocean bottom so biologists can study it.

to the type of bottom sediments; and the physical makeup of the substrate plays an important part. The firmer the bottom material, the more epifauna is found on it. Filter feeders more often locate on hard bottoms where fine particles of mud will not clog their feeding mechanism. When the bottom particles are so fine that they clog the filter feeders, we see a change in animal type to deposit feeders that take in the sediment and digest out the organic material.

In areas where there are hydrothermal vents, there are still different organisms that can live only near the vents because of their dependency on hydrogen sulfide, which comes out with the hot water. Moreover, there are always the predators and scavengers that eat what they can catch or find dead.

In studying the deep sea bottom, we find that 58 percent of it lies from 2,000 meters to 6,000 meters in depth, with a stable temperature of between 4°C and 1.2°C (40°F to 34°F). Some common animal forms found in these great depths are sea stars, brittle stars, sea cucumbers, glass sponges, sea spiders, sea lilies, polychaetes, crabs, and many fish. We have just begun to sample the large diversity of life in the deep sea. Each new expedition finds new species to describe. Each time a new way to sample is devised, new species are obtained. Deep sea sampling has turned up animal forms that were thought to have become extinct millions of years ago and others completely unknown to humans.

The Coelacanth fish (**Latimeria**) taken off Africa and the mollusk **Neopilana** taken off Mexico are examples of these. The deeper we go, the fewer species we seem to find. In the very deep ocean, in areas over 6,000 meters called the *hadal zone,* various expeditions have found that they only bring up about 10 percent of the number of species that they normally bring up in samples at shallower depths of 2,000 to 4,000 meters. The species recorded from below 10,000 meters number somewhere around 400, a very small figure compared to that of shallower depths. As we leave shore, or the littoral region, we find there are fewer genera of living organisms represented the deeper we go. The number of species per genus also diminishes with depth, as do the numbers of individuals of a species. This decline is generally attributed to the decreasing food supply as we get farther away from the photic zone where the food is originally produced.

Discovery of the chemosynthetic organisms of the deep-water vents adds an entirely new dimension to deep-water life. This will be an area of major biological concern as we move into the twenty-first century.

The benthic life in the ocean is as diverse as the types of microenvironments available to live in. The animal life, as is its general habit, has adapted to every challenge of the sea and has modified some of its forms to take advantage of all space. Of the benthic life that human beings use as food, the largest portion of it comes from the shallow seas mainly because of the higher cost of going deeper to harvest the animals there. Some of the benthic types used for food include crabs, lobster, sea cucumbers, sea urchins, sea weeds, oysters, clams, scallops, abalones, conchs, and many others. Corals, although not eaten, are used to make jewelry and displays. Humans utilize the benthos to a much greater extent than most of us realize, and as our knowledge and technology permits, we will undoubtedly use it to an even greater extent.

REVIEW QUESTIONS

1. How does the photic zone in a soft substrate differ from the photic zone in a hard substrate as to the type of animal life found there?
2. How do mud dwellers differ from sand dwellers?

3. What is the main ecological factor that changes the animal life in an area from filter feeders to deposit feeders?
4. Why does the number of animal types decrease at lower depths?
5. How does the commercial use of corals differ from the commercial use of most other benthic animals?

Chapter Nine

ARCTIC AND ANTARCTIC

ENVIRONMENTS

DEFINITION OF TERMS USED IN CHAPTER 9

Endemic species: Those species which are only found in one area are said to be endemic to that area.
Krill: A type of euphausiid crustacean, resembling a true shrimp, that serves as the main source of food for the filter-feeding whales.

The Arctic and Antarctic are similar in being at such an angle from the sun that little heat is generated there by the sun's rays. They are therefore covered by ice. While this ice is extremely significant for the land organisms, it is not so significant to marine life. The marine environment is a very rich one. The cold polar waters are rich in oxygen and nutrients and produce rich planktonic life. The water chilled in the Arctic and Antarctic sinks because of its increased density and slowly moves away from the polar regions along the bottom of the ocean basins. In the Atlantic basin these cold bottom waters actually cross the equator and a small portion of them make it all the way to the opposite pole. Given enough time, any given drop of water could visit all oceans in its travels.

The Arctic and Antarctic have different water circulation patterns, mainly because of the land masses. The Antarctic Ocean is unique in that it surrounds a continent, while all other oceans are surrounded by continents. Instead of the water circulating within a basin, it slowly circulates around the Antarctic land mass. This **west wind drift,** as it is called, is the only current that is circumpolar in nature. The Arctic Ocean is exactly the opposite. It is not a large sea and is well delineated by land masses. It is covered, for the most part, by sea ice about 2.5 meters thick because the cold air temperatures freeze its exposed surface water. Once the ice cap is formed, it insulates against the cold air temperature and protects the water from further freezing. When the sea ice breaks up, icebergs float out into the extreme northern portion of the Atlantic. These icebergs take approximately two years to melt. In contrast, the Antarctic produces much larger icebergs. The observed life of the southern iceberg is approximately ten years. These large Antarctic bergs are formed from ice that has had the opportunity to accumulate on land; consequently, they are much thicker than those created by the freezing of the ocean surface waters or formed by glacial action. A large berg in the Arctic would be perhaps 9 meters thick. In the Antarctic it might be 300 meters thick.

Although the life forms are different at the two polar regions of the

Figure 9-1 Scientists brave the worst of weather to study the sea. Even the intense cold of the Arctic does not stop their study.

earth, there are some similarities in general ecological design. In both areas, the number of types or species is reduced to a relatively small number. Many of the species found in polar seas are **endemic** species; that is, found there and nowhere else. Any plant or animal found in one area only is said to be endemic to that area. The percentages of endemic types found at the poles are higher than in most other regions, in part because of the harshness of the environment in terms of temperature, density, viscosity and low salinity. These and other factors make it difficult for some forms to survive in the region. Having fewer species, however, does not mean having less life in the water. Often when the number of species is reduced, the number of individuals of each species is increased until the biotic potential of the environment is reached.

The Antarctic Sea, because of the arrangement of the land mass, has strong upwellings that are always present in specific areas and bring rich nutrient water to the surface. This water, combined with long periods of sunlight during the Antarctic summer, creates an extremely high biotic potential for the plankton. This large biomass of plankton creates a feeding ground for larger animals, many of which feed on the plankton directly. Direct feeding on plankton, which the filter-feeding whales do, does not waste any energy through intermediate life forms and thus can support a larger number of the larger species. This is one reason why historically the Antarctic has been a major source for the whaling industry. The main whale food is a single planktonic organism called the krill, a euphausiid shrimp that grows to a size of 5 centimeters or more. Krill are very high in protein and occur in dense schools, the perfect food for the larger filter feeders as well as for many sea birds. One whale can eat as much as 6,000 pounds a day. This dense food supply also supports the Adélie penguins, terns, and other birds as well as fish, squid, and the crabeater seal. The available food supply, the slow metabolism of the cold-blooded fish, and the larger amounts of oxygen available from the cold water are probably factors in the evolutionary development of the **channichthyid** fishes. This family of fish are the only living adult vertebrates that do not have any significant oxygen carrying pigment in their blood. Their blood is clear and carries about the same amount of oxygen that ours would if we took away all the hemoglobin and left just the clear plasma. This fish is a bottom fish and is a predator. In a warmer climate, where metabolism would be higher and oxygen needed at a faster rate, this fish could not survive.

The Antarctic is one of the richest marine environments in the world. Its number of species, like the Arctic, is low, generally speaking, but its **biomass** is high. The Antarctic is one of the more ancient seas, being stable for a longer period of time than the Arctic. It has developed a more complete flora and fauna. The animals and plants commonly found here can stand temperatures down to 0°C with little or no damage, but they cannot stand

Figure 9–2   A hole through ice must be cut large enough to lower traps, other sampling devices and enough line to reach bottom. Here the line is on a large spool, and a tripod has been placed over the hole to guide the line through the center.

Figure 9–3   One must take great care not to fall into the water when working around an ice hole. A person can survive only a few minutes in the cold water, and it is very difficult to climb out.

to have their tissue frozen solid. At least one fish and several invertebrates have been found frozen in blocks of ice but still very much alive. The secret is that their tissues were not frozen. Because of the lower freezing temperature of the blood and the tissues caused by their high salt content and other metabolic materials, the fish survive. The saltier the solution, the lower the freezing temperature. In fresh water, it is 0°C (32°F); in the salt water of the ocean, about −2.5°C (27.5°F). The body fluids of some fish are saltier than the ocean water, particularly in the polar regions where there may be much fresh water dilution from melting ice. The water freezes around −1.7°C (29°F), and the fishes' tissue fluid and blood around −2.8°C (27°F). The ice around the fish acts as insulation and actually protects it from colder temperatures. This is also true of many of the high Zone I and Zone II benthic organisms.

One portion of the terrain that is greatly affected by the polar environments is the shoreline. The same area that in other regions is so rich in life, the intertidal zones, is generally rather barren by comparison in the polar regions. The ice forms along the shoreline, where there is a significant tidal change at the high tide mark. It forms what is called a **foot,** which protects the organisms that can survive being iced in from the crushing effect of the moving ice. Moving ice along the shore, whether the movement be from tidal rise and fall, currents, or pressure caused by accumulating ice, grinds off whatever life is trying to survive there. This grinding and scouring can go as deep as 10 meters, and only the animals in sheltered cracks survive. Between times of ice scouring, the life comes back to the rocky shoreline in great proliferation, but rarely does it gain maturity before the next winter and more ice scouring.

REVIEW QUESTIONS

1. What is the significance of the sun in the polar regions? Does it have more effect on the land or the sea?
2. What is the difference in the water circulation at the two poles?
3. Which pole has more ice in the water? How do you account for this?
4. How do you account for the large amount of life in the Antarctic?
5. How does the ice "foot" protect the shore animals?

Figure 10-1 Some of the seaweeds are small and go unnoticed by the average beach user like this clump of *Colpomenia*. This genus can be found on the Florida and Pacific coasts.

Chapter Ten

Tropical Environments

DEFINITION OF TERMS USED IN CHAPTER 10

Anthozoans: Cnidarians belonging to the class Anthozoa; the corals and sea anemones.
Zooxanthellae: Algae of microscopic size, which live in the tissue of many types of corals and aid them by production of oxygen.

In a great many people, there lies a dream of some day sailing a boat into a cove of a tropical island with beautiful sand beaches, palm trees, clear warm water, and coral reefs. For most of us, this mental picture has been created by the movies, television, and travel agencies. Even so, it is still typical of many of the tropical shores. There are two main life forms associated with tropical shores as indicator organisms for that region. The most outstanding and well-known are the **corals,** which are cnideria. The second is a plant, the **mangrove.**

CORALS

The corals are the best known feature of the tropical shores. In no other area of the sea does the great variety of life exist as it does on a coral reef. Coral reefs are classified into three main categories: the **fringing reef,** the **barrier reef,** and the **atoll.** These reefs are composed mainly of the stony corals, although other components are also involved. These reef-building stony corals are found in water where the temperature does not get colder than 20°C (68°F); in some areas within the tropical zone, they are inhibited by an influx of cold water. The cold water may come from a current, such as the Peru current off Chile in South America, or from deep water upwellings that occur regularly because the winds blowing steadily offshore push the warm surface water out to sea, as they do off the west coast of Australia.

These anthozoans are in many ways similar to the sea anemones. The most obvious difference is the hard calcium carbonate shell they secrete around themselves. They have the ability to reproduce by budding and, consequently, to form colonies of many individuals. These colonies are what form the reefs. Each species forms a colony of a different shape, hence the great variation in the reef formation. In the Indo-Pacific, the richest coral area in the world, there are approximately 700 species of coral. Although no warmer than many other areas at present, the Indo-Pacific has been a warm sea for a longer period of time than most other areas because of its isolation from major influences during the ice ages. This longer period of warm water has allowed many more modifications to occur and, consequently, a larger number of species to develop.

As mentioned before, the reefs are not composed purely of coral skeletons but have many other components. These other components include shells of various mollusks, debris that washes on to the reef and becomes lodged there, and very significant portions of coral-like algae. All coral reefs have a large amount of microscopic algae growing within the tissue of the coral polyp. These algaes are called *zooxanthellae.* These small units of algae have been studied extensively and seem to have several functions that enable the coral to exist in such dense colonies. The zooxanthellae produce oxygen through photosynthesis, use up some of the waste products of the coral polyp, and produce glycerol for the coral polyp to use as a nutrient to supplement its own filter feeding process. The reefs cannot exist without this symbiotic zooxanthellae presence. Although the most important aspect of this relationship is unsure, it could well be the assimilation of toxic waste material given off by the dense colonies of coral polyps. It could also be, as some believe, the production of oxygen for use by the coral when the water is low, calm and warm, times of low oxygen content in the water. When sufficient research has been completed, it will probably show that the com-

Figure 10-2 Scientific divers use a magnifying glass to study the small coral organisms.

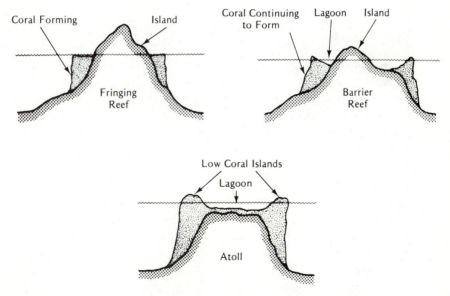

The coral continues to grow as the island is eroded until only the coral is visible.

Figure 10-3 Formation of a coral reef and island.

bination of oxygen reproduction, waste disposal, and food production, rather than a single factor, fosters growth of these dense colonies of coral.

We mentioned above three types of reefs, fringing, barrier, and atoll. We can describe how these reefs differ by considering a classic example of an island created by a volcanic eruption. The island is formed by volcanic activity. Where the conditions are suitable for coral growth, coral will start to accumulate in the shallow water around the islands. Coral can create reefs only in shallow water where their zooxanthellae can get sunlight. This coral growth would be called a fringing reef because it creates a fringe around the island. As the island subsides, the original edge of the reef becomes farther from the land mass of the island. The coral on the inside of the reef or in the lagoon that is formed does not thrive so well as that on the outside of the reef. This is in part due to less oxygenated water on the protected side of the reef and less plankton to feed on, because the plankton was filtered out as it came across the reef. Different species that are better adapted to the lagoon environment become evident. The lagoon becomes larger as the land sinks until at some point the reef is far enough away from the shore to be termed a barrier reef. The most famous of all coral reefs, the Australian Barrier Reef, is some 1,200 miles long and is composed in reality

Figure 10-4   A colony of sponges creates a microenvironment for small fish, crabs, and many other organisms.

of many different types of reefs. In our example, if the island continues to erode and/or sink, it will disappear entirely, leaving a series of small coral islands in a somewhat circular pattern with a lagoon in the center. These islands are called atolls. Some atolls are quite large, and some lagoons are over 80 kilometers (50 miles) across. The South Pacific, due to its high volcanic activity in the past ages, has thousands of atolls that were created in just this manner.

Because of the great variety of life forms and the bright colors found on many of the organisms living in and around the coral reefs of the world, they have been photographed and written about more than any other group of marine animals. The coral reef is paradise for the underwater photographer. We see and read so much more about the tropical environment than we do any other marine region that we tend to think it is the most important. This is not the case. As beautiful and awe-inspiring as the tropical region is, the fisheries of the world and, consequently, the oceanic food supply depend mainly on colder and less publicized regions. This fact is not meant to reduce the importance of the tropical regions of the world in any way, but rather gives the proper prospective as to their place and significance in the marine environment as a whole.

MANGROVES

While corals are the most characteristic animal life of the tropics, the mangrove is the most characteristic plant. The mangrove is able to grow on most types of shore substrates. Trees can get quite tall, around 6 meters, but in most mangrove areas are only 3 to 4 meters high. Their ability to grow in salt water is their main physiological adaptation to their environment. They grow in thick fringes along many tropical shores and particularly in shallow bays and protected areas, creating the mangrove swamps of the tropical world. The most obvious physical feature different from most other trees and bushes is the way the roots show above ground. Called *strut roots,* they start out of the trunk above the ground, spread out and then into the ground. This type of root gives a firm base to the mangrove tree so the water will not erode its base or the winds blow it over.

When examining the base of the mangrove tree, one finds many similarities with the holdfast of the larger forms of brown algaes. The many fingerlike projections giving support and holding strength appear similar to the eye, even though they are not. The mangrove roots penetrate the ground and draw nourishment from it. The algae holdfast does not penetrate the substrate and does not draw nourishment from it. The physical appearance,

Figure 10–5   This flamingo tongue shows how it feeds on the polyps of a soft coral. The bare portion around it has all the organisms eaten away.

Figure 10–6   The squirrelfish is a common coral reef fish. Benthic in nature, it is rarely seen off the bottom.

Figure 10–7 There is so much to see on a coral reef that one could spend an entire lifetime observing and still see but a portion of what is there.

Figure 10–8 Many forms of smaller plants move onto the deltas created by the mangroves, and eventually a characteristic biological community becomes stable. This is one of the climax communties of tropical regions.

Figure 10–9 Little by little, silting creates a dry delta that moves slowly outward into the bay.

however, is similar, and so is the use made of these parts by the other organisms commonly living in these environments. The area protected under a kelp holdfast is alive with many small life forms. The area protected by the many branches of the strut root of the mangrove is also teeming with small life forms that use it as protection, the most commonly seen of these being one or another species of crab. The swimming crab can often be seen darting for safety under the root cover of a mangrove. Also common is the oyster, which grows on the exposed roots. The mangroves grow in the shore zones we have already described in earlier chapters. In Zone I, they start; Zone II is their stronghold; and they rarely go deep enough to be out of Zone III. Their presence, however, does not preclude the worldwide distribution of barnacles in Zone II. The barnacle is still an indicator organism for Zone II areas where conditions are available for their growth.

REVIEW QUESTIONS

1. What are the three main types of coral reef? How do they differ from each other?
2. What is the major limiting factor for coral reef formation?

3. What is the significance of the zooxanthellae?
4. What is the main adaptation for mangroves to grow in any given environment?
5. How does an algae holdfast differ from the roots of land plants?

Chapter Eleven

NORTH AMERICAN

MARINE ENVIRONMENT

DEFINITION OF TERMS USED IN CHAPTER 11

Subtidal biology: A term used to designate the study of organisms below the lowest low tide.

Transition area: Where two zones come together and species from each zone overlap.

North America has a variety of environments because it divides two oceans and extends from the equator north to the Arctic in both of them.

CURRENTS

On the East Coast the North Atlantic equatorial current is driven by the rotation of the earth against land around Cuba and Florida then is deflected up the Florida coast, where it is joined by other waters from the Gulf of Mexico and the Caribbean to form the **Gulf Stream.** The Gulf Stream is a rapidly moving current, for the most part, and is warm water. Comprised of many different smaller currents, it is really a current "system" rather than a single entity; and because of its complex makeup, it is variable in speed, size, and direction. This warm water is of major influence all the way up the Atlantic coast of the United States. The Labrador current, which flows southward to meet the Gulf Stream around Newfoundland, brings down cold water that mixes with the coastal waters of the North Atlantic states. This cooling action creates an entirely different fauna than is present farther south along Florida.

In the Pacific there is a similar circulation, with the **north equatorial current** being driven to the coast of Asia and turning north to become the **Japanese current.** The Japanese current is not so dramatic a current as the Gulf Stream but resembles it in its general circulation. It travels north and east, with part of it nearly reaching the Hawaiian Islands. The northern portion of the current is joined by a cold, south-flowing current, the Oyashio. It then travels on to Alaska and branches again to join with the North Pacific current and becomes the **California current.** The California current runs down the Pacific coast and starts to swing out to sea in Southern California to rejoin the north equatorial current on its way to Asia. The California current is subject to water from upwelling along the coast as it travels south. This cold water helps to drop the temperature of the current to make it a fairly cold body of water, with a temperature range approximately 11°C (52°F) to 21°C (70°F).

TEMPERATURE

Because of the relatively small geographical temperature differences along the Pacific coast, the territorial ranges of many of the animals are unusually extensive. *Pisaster* is a common genus of sea star from Alaska to the tip of Baja California. Many forms which live in shallow water in northern sections of the range along British Columbia and Washington are also found in deeper water off Southern California. They seem to stay within a range of colder water. The surface waters in the Pacific Northwest are about the same temperature as the water off Southern California at a depth of 35 to 65 meters. With the advent of "**subtidal**" biology, which has become common only since the development of scuba gear, the ranges of many species have been found to be much greater than previously estimated.

REGIONS

Many animals thought to be local in a given area have been found living some distance away at a deeper depth. When a close inspection of the two coastlines is made, both the Atlantic and the Pacific coasts can be divided into three general regions. On the Pacific coast the north region is above or north of Vancouver Island in British Columbia. The central region is the area from Vancouver Island in Canada south to Point Conception in the southern third of California. The southern region extends from Point Conception to the Cabo Falso at the tip of Baja California in Mexico. On the Atlantic coast the northern region lies north of Cape Cod, the central region from Cape Cod south to about Cape Canaveral, and the southern region from Cape Canaveral into the Gulf of Mexico.

It is interesting to note that in the transition area where these regions come together, there is an overlap of species from both regions. Quite often different species that have short ranges and do not usually extend successfully into either zone are also present. This variety makes the transitional areas extremely interesting to the biologist. The author found data easily available from all of the main transitional areas around the world, but rather hard to obtain about the center of a region. When asked to comment, one researcher replied, "Why should I work in a stable, rather boring environment, when I can work in a changing, exciting one." This response should remind us that scientists, regardless of how dedicated, are still human.

The northern regions on both coasts have species with long ranges extending into the region or regions next to them, but the Atlantic coast has fewer long-range species than the West Coast. Although many Atlantic forms extend into the Arctic, the large majority range only within their own region. On the Pacific coast, more forms extend into the central region than into

Figure 11-1 The lighthouses of years ago are slowly disappearing, and electronic devices are taking their place.

the Arctic. The line separating the northern and central regions on the West Coast is not so well defined as it is on the East Coast.

On the East Coast more species from the southern region extend into the central region, whereas the West Coast central region has more in common with the northern region than the southern one. This difference can partially be explained by the direction of water flow along the coasts. On the East Coast the general current flows north; this effectively prevents the planktonic species from extending their range to the south. On the West Coast, the general water flow is to the south, tending to inhibit northward extension of life forms. Because benthic forms have a better chance of extending their range than do pelagic forms against currents, we would expect, at least in theory, that the benthic forms would be more widespread than the pelagic ones. Investigation shows that there are, in fact, more species of the benthic variety with extended ranges than there are pelagic species.

The southern regions on both coasts have most distinctive faunas. Southern California below Point Conception is not in any way a tropical environment; only a few of the tropical forms from the Gulf of California extend more than a third the way up the Baja California coast. It is considered a warm temperature environment, in contrast to the southern region on the East Coast where the majority of species are tropical in nature. Along the Florida Keys, the fine silt settling out of the water is probably also significant in controlling certain species that might otherwise do well but are not found here because of the smothering effect of these fine sediments.

Southern Florida and the Keys, which extend south and west from it, create a well-defined area. The land extends over a hundred miles offshore on the west side of Florida at 100 meters or less in depth. Along the southern portion of this very shallow Florida plateau, high ground extends above the water and forms many small islands called the Florida Keys. Between the Keys and the coast, the water in most places is less than 4 meters deep, although it may be 30 kilometers from land. It is in this area, so far from land and yet so shallow, that many unsuspecting sailing ships of the past met their doom. These warm, shallow waters contain many corals and gorgonians. The gorgonians stand tall off the bottom and bend with the currents to give the illusion of forests bending in the wind. They are quite spectacular. The farther eastward away from the silt of the Gulf of Mexico we investigate, the more coral we find. The main genera of this area are **Acropora, Montastrea, Siderastrea, Porites,** and **Manicini.** These forms are less affected by the silting and temperature changes and do well here.

The zones we have discussed in earlier chapters also exist here. The isopod *Ligia* is found on the Keys in Zone I, just as it is in Zone I of the California coast. *Littorina* is also found here in this zone, as it is in most parts of the world. A dark colored area at the lower portion of Zone I, like the ring around a bath tub where the edge of the water touches the tub, is

Figure 11–2   Shore crabs are among great performers of rocky shore regions.

produced mainly by the dark-colored algae, *Entophysalis, Brachytrichia,* and *Tellamia.* Zone II has the barnacle, *Chthamalus,* and the alga *Valonia* as the common, widespread forms. In Zone III, we find *Gelidium* and *Laurencia,* as two algae characteristic of this zone on rocky areas. *Mytilus* and *Pachygrapsus* represent a bivalve and an arthropod that are widespread in Zone III on both the Atlantic and Pacific coasts. Zone IV has some coral, but not a great deal. The mussels of the genus *Arca* are often found here, as is the calcareous alga *Halimeda.* Many forms call this zone their home; but, as we noted before, it is so varied in nature because it is normally under water that it is the least definable of all zones.

As we travel along the coast, we find in northern Florida a coastline that receives a steady flow of breakers and sand scouring, in contrast to the little wave action and silting in the south. One of the few animals that can stand so drastic a change is the Zone I *Ligia.* Of course, it is above the wave action and is more affected by air temperature than other factors. The main boundary seems to be at Cape Canaveral, a winter temperature barrier; few of the tropical species range above that point.

Farther up the Atlantic coast near Beaufort, North Carolina, the temperature ranges from 10°C (50°F) to 27°C (80°F), at nearby Cape Lookout. Here are reefs composed not of rock but of peat, an organic material partially carbonized by decomposition in water. We find *Ligia* still present, along with a crab, *Uca,* in Zone I. The mid-tide Zones II and III are characterized by *Chthamalus* and *Balanus* in Zone II and *Crassostrea* and *Mytilus* in Zone III. The algae *Ulva* and *Enteromorpha* are also common, as they are in the same zone of the southern region on the West Coast. The most conspicuous life forms of Zone IV are the algae; this area is one of the richest in algae production. The thick algae create cover for many animals; consequently, the area is a very exciting one biologically. This region has many similarities to northern Baja California and Southern California on the West Coast.

The northern Atlantic region has colder water. Off Nova Scotia surface temperatures range from 5°C (41°F) to about 19°C (66°F). The cold water becomes the barrier for the southern species. The intertidal forms may have to endure 0°C (32°F) temperatures to survive. In winter, ice quite often forms along protected shores. The weather is cold for *Ligia;* it is rarer here than in warm areas and at times not found in all. *Littorina* is still characteristic of Zone I, however, and extends down into Zone II. Zone II is again characterized by its barnacles. In this area *Balanus* is the main genus. Zone III is rich with the brown algae, *Fucus* and *Ascophyllum. Fucus* is well adapted to this area in having several species that are modified to exist in slightly different environments, depending on exposure to air. The Zone IV dominant type is the alga *Laminaria.* Very well adapted and out-competing some other forms for space, it is by far the most dominant life form of this zone.

TABLE 11-1 COMMON ORGANISMS OF THE NORTH AMERICAN MARINE ENVIRONMENT

Algae	
1 *Valonia*	
2 *Gelidium*	
3 *Laurencia*	
4 *Halimeda*	
5 *Ulva*	
6 *Enteromorpha*	
7 *Fucus*	
8 *Laminaria*	
9 *Nereocystis*	
10 *Egregia*	
11 *Endocladia*	
12 *Macrocystis*	
13 *Eisenia*	
14 *Postelsia*	
15 *Entophysalis*	common ''Black'' line in Zone 1
16 *Brachytrichia*	
17 *Tellamia*	

TABLE 11–1 *(continued)*

Cnideria	

1 *Acropora*
2 *Montastrea*
3 *Siderastea*   corals
4 *Porites*
5 *Manicina*
6 *Anthopleura*   large green anemone
7 *Gorgonian*   order name for the sea fans

Arthropods	

1 *Chthamalus*
2 *Balanus*   barnacles
3 *Tetraclitia*
4 *Pollicipes*
5 *Pachygrapsus*—shore crab
6 *Uca*—fiddler crab
7 *Ligia*—most common Zone I isopod

Mollusca	

1 *Littorina*—most common Zone 1
2 *Tegula*—black turban   gastropods
3 *Haliotis*—abalone
4 *Acmaea*—limpet
5 *Dendropoma*—a tube snail
6 *Mytilus*—most common mussel   pelecypods
7 *Arca*—a common bivalve
8 *Tivela*—pismo clam
9 *Crassostrea*—oyster

Echinoderm	

1 *Pisaster*—common sea star

The Pacific coast is less distinct in its zones because its temperature range is narrower than that of the East Coast, so species can more easily move into other regions. The west coast of Canada is protected in part by Vancouver Island. The island protects about 400 kilometers of coast, or, because of the many inlets, 3,000 kilometers of shoreline. The water circulation around the island is very poor because of the small islands between the big island and the mainland about 160 kilometers south of the northern tip of the island. For a distance of around 120 kilometers, the passage through the channel is narrow and broken by many islands that inhibit water flow. Off the southeast end of the island, there is a similar situation; the many islands in the path of water flow slow down the circulation considerably, even though the main channel does not narrow in this area. The water enters the passage around Vancouver Island from the north through Queen Charlotte Strait, and from the south through the Straits of Juan de Fuca. Between the mainland and the island and between the north and south restrictions in the main channel is the Strait of Georgia, a deeper body of water than either entrance. The Strait of Georgia, approximately 230 kilometers long, contains relatively few islands and, because of its more isolated nature, has temperatures as warm as 18°C (65°F). Directly across the island, perhaps 64 kilometers distant, the warmest temperature is around 13°C (55°F).

Other interesting things happen in the Georgia Strait, such as the high fluctuation of the salinity. Normally it is lower than in the open sea, as might be expected due to the great fresh water runoff from the rain. The most runoff, and therefore the lowest salinity, generally occurs in the summer. The salinity varies from 33 percent to as little as 21.5 percent. Vancouver Island is a unique environment because of its great diversity, not only in

TABLE 11–2 SOME COMMON FORMS BY TIDAL ZONES AND NORTH AMERICAN REGIONS

	North Pacific	North Atlantic
I	Ligia—Littorina	Littorina
II	Balanus	Balanus
III	Chthamalus—Mytilus—Fucus	Fucus—Mytilus
IV	Pisaster—many algae	Laminaria

	Central Pacific	Central Atlantic
I	Liga—Littorina—Black line algae	Ligia—Uca—Littorina
II	Balanus	Chthamalus—Balanus
III	Ulva—Enteromorpha—Tegula	Crassostrea—Mytilus—Ulva—Enteromorpha
IV	Haliotis—Egregia—Macrocystis	Many algae

	South Pacific	South Atlantic
I	Ligia—Littorina	Ligia—Littornia—Black line algae
II	Acmaea—Balanus—Chthamalus	Chthamalus—Valonia
III	Tetraclitia—Mytilus—Pollicipes—Anthopleura	Gelidium—Laurencia—Mytilus—Pachygrapsus—Arca—Halimeda
IV	Many algae—Tivela	Gorgonians—corals

Figure 11–3 The gooseneck barnacle is common in the lower tidal zones and widespread geographically because it sometimes grows on drift wood and is carried by currents to all parts of the oceans.

possibilities for microenvironments between its west and east coasts, but also in physical factors such as temperature and salinity. In its protected waters we find the same indicator organisms as we did on the Atlantic coast in Zone I, mainly the isopod, *Ligia*. *Littorina* is also very dense in population here, as in almost all Zone I areas of the West Coast environments. Zone II is distinctly a barnacle zone. The two main genera are *Balanus,* with several species found in the lower part of Zone II and on into Zone III. *Chthamalus,* found mainly in Zone III, has not the zonal range that *Balanus* has. Zone III is delineated with the algae *Fucus* and the mussel *Mytilus*. Zone IV varies greatly in the area due to the great density of environmental factors, but in general *Pisaster* is the conspicuous, dominant genus. Many algaes are found here, but none are the single, characteristic Zone IV type. They vary greatly from location to location. The large kelps *Nereocystis* and *Egregia* are dramatically present and impressive in their size. The temperature in this region is one in which most algaes do well. *Ulva,* as an example, which in the southern region generally does not grow to more than a few centimeters across, in the northern region often exceeds 60 centimeters. *Nereocystis,* a brown algae, is the most striking form, reaching over 30 meters in length.

The central region, of which northern California and Oregon are a part, is noted for its open coastlines with heavy surf pounding in from storms far

Figure 11–4 The *pneumatocyst* or flotation bladders are plainly visible on *Macrocystis* as it grows off the West Coast of the U.S.

out at sea. Zone I is easily found by its indicator organisms, *Ligia,* and the mixed algae growth that creates the black line at the upper edge of the sea.

Zone II, as is characteristic of areas with a combination of quiet water in bays and rough water on the open coast, has various degrees of sharpness in its delineation. Where the surf is large and the spray and surge sweep up and down, there can be no sharp line between the zones. Transition areas become the rule and can be confusing to a student looking for the well-defined zones described in textbooks. To receive good training in zone recognition, it is best to start in a bay where the quiet water allows for sharp zonal differentiations. After one has learned to recognize the features of each zone in the bay, they become much easier to distinguish in the open coast area. The main indicator organism here is *Balanus.* Because so much of this central region is exposed and the zones more often than not are poorly delineated, we will find many forms that are more at home or under less environmental stress in Zone III ranging up into Zone II. A good example is the algae *Endocladia.* In rough areas it grows well into Zone II; in quiet waters it does not. The black turban snail, *Tegula,* is another example of a poorly delineated zonal species. It occurs in Zones II, III, and IV, but its densest population is in Zone III. Consequently, one who merely looks for its presence could be misled. It becomes necessary to make either a subjective or an objective population density count before the presence of this organism becomes significant to zonation.

Zone III contains, as always, a great variety of life. The mussel *Mytilus* becomes present in this zone and extends into Zone IV, as does the abalone *Haliotis*. Abalone, once very common both in shallow and deep water, has been nearly exterminated in some areas by a combination of the commercial abalone diver, the sports diver and the sea otter. Laws to protect the abalone are now in force. It is unfortunate that sea otters cannot read, because in the areas where they range, the abalone has been exterminated.

Zone IV is heavy with laminarians and various red algae. *Egregia* is one of the larger inshore algae reaching 5 meters. The large kelp *Macrocystis* forms thick mats on the surface beyond the breaker line and forms the characteristic kelp beds of the central and southern regions.

In the southern region, extensive sand beaches occur mainly in the northern sector and dominate several hundred miles of coastline. Farther south are mainly rocky coasts, with sand bays along the central Baja California coast. Although most of the area is exposed to open sea and strong wave action, it is farther south and farther from the strong storm centers of the North Pacific. The waves here are generally smaller than along the central and northern exposed areas.

Zone I is characterized by the forms that are normal for it, with one modification. Much of the rock of this upper zone is soft limestone and holds moisture very well between wettings. This allows some animals to range farther up into this zone than we normally find elsewhere, particularly in the San Diego area. The sandstone is not the best substrate for barnacles because it is easily eroded, so that barnacle population is down and smaller barnacles rather than large ones are more prevalent. *Littorina* is the dominant form.

In Zone II the limpet *Acmaea* is common along with the barnacles *Chthalamus* and *Balanus*. The larger red barnacle, *Tetraclitia* is transitional between Zone II and III, occurring in greater numbers in Zone III.

In Zone III the dominant form is the mussel *Mytilus*. The stalked or goosenecked barnacle *Pollicipes* is also almost always present. In tide pools, the anemone *Anthopleura* is conspicuous.

Zone IV has the normal algae growth, with one of the easily recognizable forms being the small palm kelp *Eisenia*. Appearing like a 2-foot palm tree, it is similar to *Postelsia* of the central region, but has wider leaf-like fronds.

A rather famous inhabitant of the sandy environment in Zone IV is the pismo clam *Tivela*. At very low tides in some areas, people arrive at the beaches with pitchforks to probe into the sand and locate the pismos. They are excellent eating.

REVIEW QUESTIONS

1. What is the major difference between the Gulf Stream current and the California current, in the effect they have on the distribution of animal and plant life along their prospective coastlines?

2. Why do biologists prefer to study the areas where two zones come together?

3. Zone I can be recognized in most parts of the world by the presence of several indicator organisms. What are they?

4. Why is it easier for species to move from one zone to another on the Pacific coast than on the Atlantic coast?

5. Why is it best to start the study of zones in a bay?

Part Three

LIFE IN THE MARINE ENVIRONMENTS

Figure 12-1 The avocet can be identified by its upturned bill. It can be found from the Gulf of Mexico to the Pacific coast.

Chapter Twelve

A SYSTEM OF NAMING

THE MAIN GROUPS

DEFINITION OF TERMS USED IN CHAPTER 12

Analogy: Correspondence in function but not in origin; (for example, the wing of a bee and the wing of a hummingbird).

Binomial nomenclature: The system using two names, genus and species, to identify organisms.

Homology: Correspondence in structure or different organisms that seem to have the same evolutionary origin; (for example, the leg of a horse and the leg of a cow).

Linnaeus: The scientist who established the foundation of our modern system of scientific naming in the early 1700s.

Saprobic: An organism that derives its nutrition from nonliving organic material.

Taxonomy: The study of the naming of organisms.

The naming of the things around us became necessary when humans developed a spoken language. To refer to something that is not present to point at, one must use some word or name that is recognized by the listener. Thus, through necessity, came the names that we struggle with today. Each known organism has a distinct individual name, a scientific name. The same scientific name applies to all organisms that have the same characteristics.

One of the first attempts to name, and therefore to classify, animals by groups was made by **Aristotle.** He grouped all animals into those that have blood and those that do not, but unfortunately he did not recognize blood unless it had a red pigment. Many types of blood do not have red pigments; they may be clear fluids. It was not until the early eighteenth century that accurate classifications really began with the naming system devised by **_Linnaeus._**

Linnaeus believed that with no more than a dozen words, it was possible to differentiate every living organism from every other living organism. To accomplish this the observer must be able to "see" and "recognize" the important differences in each organism. This takes great skill, knowledge, and experience. Although Linnaeus was a professor of botany at a university in Sweden, his naming system works as well for animals as it does for plants. Linnaeus's system was improved when Lamark and Swier started to differentiate life forms according to the basic plan of tissue organization in their bodies. This principle founded a system of determining one type from another that worked and is still used today.

Basically, those organisms that have similar anatomy and physiology are placed in one group. Then types within that group with similar traits are placed in a subgroup, and so on until we have the last subgroup that has the most minutely detailed definition of characteristics. This lowest group is called the **species.** A species is generally a group or population capable of interbreeding. In some species minor characteristics further divide the species into subspecies or races, but the species is generally considered the end of the subdivisions. The grouping directly above the species level is the **genus.**

Because many organisms have the same species name, just as many people are named John, the genus name is used along with the species name, just as a first and last name of a person are used to identify that per-

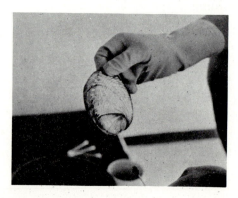

Figure 12–2 This abalone shell has been very carefully cleaned with acid to remove any growth on it. Now it can be studied to see the features that characterize its species.

son. The genus name is written first and is always capitalized and underlined or printed·in italics. The species name is next and is never capitalized, but it also is underlined or written in italics. An example would be *Octupus bimaculoides, Octopus* being the genus name and *bimaculoides,* the species name. The system, although simple in principle, has become complex due to the number of things that must be named. The science of naming is called *nomenclature* and is a part of **taxonomy,** the science of classification and identification. This area of science requires specialists just like any other special-interest areas in science. Taxonomists concern themselves with characteristics and make decisions as to whether or not they show **homology** (similarity of origin) or **analogy** (similarity of function only). For example, the claw of a crab and the mechanical claw of a piece of construction equipment are analogous but are not homologous. Based, then, on homologous characteristics, the animals and plants are classified and named. Because these names and groups are made by humans, they are not perfect and undergo change from time to time as new data are obtained. At times arguments even arise about who has classified and named something properly. If in your reading and reference work you find two books calling the same organism by different names, don't be concerned. The organism isn't.

This system of giving two names, the genus and species, to each organism is referred to as ***binomial nomenclature.*** It has been the standard method of naming since 1758 when Linnaeus published his tenth edition of *Systema Naturae.* In 1898 the International Congress of Zoologists established an ongoing Commission for Nomenclature, which prepared an International Code of Nomenclature, and acts as judge in deciding the appropri-

Figure 12–3 Small organisms like this tiny cling fish hiding on the piece of kelp are difficult to find initially, but may be easy to identify because their camouflage coloring makes them unique.

ate names if a conflict arises between two proposed names of the same organism. The code gives the procedure of naming from family down to subspecies. It read as follows:

1. Zoological and botanical names are distinct (the same genus and species name may be used, but it is not recommended, for both an animal and a plant);

2. No two genera in the animal kingdom may bear the same name, and the same applies to two species in a genus;

3. No names are recognized prior to those included by Linnaeus in the *Systema Naturae,* tenth edition, 1758;

4. Scientific names must be either Latin or latinized and preferably are printed in italics;

5. The genus name should be a single word (nominative singular) and begin with a capital letter;

6. The species name should be a single or compound word beginning with a small letter (usually an adjective agreeing grammatically with the genus name);

7. The author of a scientific name is the person who first publishes it in a generally accessible book or periodical, with a recognizable description of the animal;

8. When a new genus is proposed, the type species should be indicated;

9. A family name is formed by adding IDAE to the stem of the name of the type genus, and a subfamily name by INAE.

The major groups of animals are called *phyla* (singular, *phylum*). They are characterized on Table 12–1. The major or more common phyla, which will be treated in detail, with an entire chapter devoted to each one, are the Protozoa, Porifera, Cnideria, Arthropoda, Echinodermata, and Chordata. The miscellaneous phyla or those of less interest at this time to the student will all be discussed in one chapter.

A basic and incomplete breakdown of some of the more important groups included in the scope of this book is included in the following Table:

TABLE 12–1

Kingdom Monera

These forms lack nuclear membranes and most organelles associated with cells. They generally only have one chromosome.

Phylum Eubateriae	The true bacteria
Phylum Cyanophyta	The blue-green algae

TABLE 12-1 (*continued*)

Kingdom Protista

These organisms are composed of a single cell (some are colonial), which has a nucleus and multiplies through the process of mitosis.

(Plantlike groups)
Phylum Euglenophyta	Euglenoids
Phylum Chrysophyta	Golden-brown algae
	Yellow-green algae
	Diatoms
Phylum Pyrrophyta	Dinoflagellates

(Animallike groups)
Phylum Protozoa	Unicellular microscopic animals
Class Flagellata	Flagelates
Class Sarcodina	Free-living amoebas
Class Sporozoa	Endoparasites
Class Ciliata	Have cilia

Kingdom Fungi (Myceteae)

These are nonphotosynthetic organisms. They are saprobic or parasitic plants that absorb nutrients through their cell membrane after extracellular digestion.

Division I	— Mastigomycota	—	Also classified as heterotrophic protista—molds
Division II	— Amastigomycota	—	True fungi

Kingdom Plantae

These plants, normally called the "higher plants," are composed of two subkingdoms: nonvascular and vascular plants.

Subkingdom Thallophyta

This group is composed of the multicellular nonvascular algae, most of which occur in water.

Phylum Chlorophyta	Green algae
Phylum Phaeophyta	Brown algae
Phylum Rhodophyta	Red algae

Subkingdom Tracheophyta

This group is composed of the multicellular vascular plants, most of which occur on land.

Phylum Bryophyta	Liverwarts, mosses, etc.
Phylum Pteridophyta	Ferns, etc.

(*continued*)

TABLE 12–1   *(continued)*

Subkingdom Tracheophyta

Phylum Spermatophyta   Seed plants
    Class monocotyledonae
    Includes salt-water grasses such as Zostera and Phyllospadix.

    Class Dicotyledonae
    Includes the mangroves, pickleweed, and marsh rosemary.

Kingdom Animalia

These are animals that are metazoans (having more than one cell). (The following list of phyla *is not complete*. Because of the scope of this book, many of the smaller phyla or those without major marine representatives have been omitted from this list).

Phylum Porifera	Sponges—possess spicules. Mostly marine
Phylum Cnidaria (Coelenterata)	Hydroids, jellyfish, sea anemones, corals, gorgonians radial symmetry—have nemotocysts
Phylum Ctenophora	Combjellies, mostly planktomic all marine
Phylum Platyhelminthes	Flatworms—bilateral symmetry
Phylum Nemertea	Ribbonworms—mostly marine—have mouth and anus— closed system for blood
Phylum Rotifera	Rotifers—Ciliated oral disk
Phylum Nematoda	Roundworms—complete digestive tract
Phylum Bryozoa	Encrusting colonies
Phylum Brachiopoda	Lamp shells—shells open dorsally
Phylum Annelida	Segmented worms
Class Polychaeta	Most marine annelids
Phylum Echiura (Echiuroidea)	Spoon worms; live in a U-shaped tube
Phylum Sipuncula	Peanut worms, unsegmented worm with tentacles around mouth
Phylum Mollusca	Mostly with calcareous shells (sea shells); shell of 1, 2, or 8 parts
Class Amphineura	Chitons—shell of 8 parts
Class Gastropoda	Having mostly one shell; snails, etc., and nudibranchs
Class Bivalvia	Having two shells—clams
Class Scaphopoda	Tooth shells—shell of 1 part
Class Cephalopoda	Octopuses, squids, etc.
Phylum Arthropoda	Having jointed legs and a chitinous exoskeleton
Class Merostomata	Horseshoe crabs
Class Arachnida	Marine mites (land spiders)
Class Pycnogonida	Sea spiders; marine
Class Crustacea	Most marine arthropods; copepoda—crabs—shrimp— etc.
Phylum Chaetognatha	Arrow worms; hairlike bristles around mouth; marine
Phylum Echinodermata	Sea stars, sea urchins, etc.; radial symmetry; tube feet; marine
Class Asteroidea	Sea stars
Class Echinoidea	Sea urchins, sand dollars

TABLE 12–1   (continued)

	Kingdom Animalia
Class Holothuroidea	Sea cucumbers
Class Ophiuroidea	Brittle, serpent, and basket stars
Class Crinoidea	Sea lilies
Phylum Hemichordata	Acorn worms
Phylum Chordata	A single, dorsal, tubular nerve cord, gill slits in the pharynx, and a notochord at some stage in their life cycle
Subphylum Protochordata	Primitive forms
Class Urochordata	Tunicates, salps (Considered by some to be a subphylum)
Class Cephalochordata	Lancelets (Considered by some to be a subphylum)
Subphylum Vertbrata	Advanced forms with a dorsal support column
Class Agnatha	Cyclostomata
Class Chondrichthyes	Have a cartilaginous skeleton—sharks, rays, and chimaeras
Class Osteichthyes	Have a bony skeleton; most of the common fishes
Class Reptilia	Sea turtles, snakes, and lizards (many land forms)
Class Aves	Birds
Class Mammalia	Humans, whales, seals, dogs, etc.

The groups used to subdivide phyla are class, order, family, genus, and species. Sometimes more groups become evident as the organisms are studied. In these cases the term "super" or "sub" may precede one of the standard divisions, and the term "division" is also used at times. An example follows in the classification of the East Coast crab commonly called the "blue crab."

Phylum, Arthropoda

Class, Crustacea

Subclass, Malacostraca

Division, Thoracostraca

Order, Decapoda

Suborder, Brachyura

Superfamily, Brachyrhyncha

Family, Portunidae

Genus, *Callinectes*

Species, *sapidus*

TABLE 12–2

Phylum	Approximate number of species*
Arthropoda	900,000
Mollusca	50,000
Chordata	42,800
Platyhelminthes	12,700
Nematoda	12,000
Annelida	11,400
Cnidaria	9,000
Echinodermata	6,000
Porifera	5,000
Bryozoa	4,000
Rotifera	1,500
Nemertea	900
Brachiopoda	350
Sipuncula	325
Echiura	140
Ctenophora	100
Hemichordata	80
Chaetognatha	70

*Phyla of the kingdom animalia ranked according to the number of species in the phylum. The numbers are growing continuously as new species are discovered and described. We also are losing species continuously because of the ability of humans to outcompete other species.

Figure 12–4 The discovery of the remains of dead organisms can be helpful in establishing the proper classification of a group of organisms, as well as helping us to understand much of our past history.

The names of each group tend to tell what the main characteristics of that group are. An example, the same "blue crab" again, is as follows:

Arthropoda—having legs with joints
Crustacea—having an exoskeleton
Malacostraca—having a softened shell
Thoracostraca—having a thorax
Decapoda—having 10 legs
Brachyura—having a shortened tail section
Brachyrhyncha—having a shortened nose
Portunidae—mythological character, god of the harbor
Callinectes—"beautiful swimmer"
sapidus—good tasting

These words seem difficult to us because they are, for the most part, Latin. They do, however, make sense in the international taxonomic language, which is Latin. Although there is fortunately no reason to learn most of these terms, the phyla names are important and should be memorized along with the main characteristics of each one. In the major phyla one should memorize some class names and the scientific names of the most common local species, along with a few well-known international species of significance. The marine biologist must learn several hundred of these terms, but the basic student or the generally interested person need only learn a few to achieve greater enjoyment of this remarkable environment.

REVIEW QUESTIONS

1. Who were the scientists who started the system of classification we use today?
2. How are organisms classified? Give names beginning with the most general grouping to the most specific.
3. What is binomial nomenclature?
4. What is the reasoning for the scientific names given an organism?
5. What is the difference between *analogous* and *homologous* features?

Figure 13–1 The bottom trawl is dragged across the sea bottom and catches many representatives of the benthic community for study.

Chapter Thirteen

MARINE PLANTS

DEFINITION OF TERMS USED IN CHAPTER 13

Alginates: A group of compounds derived from brown algae and used as thickeners in foods and other products.

Agar: A compound derived from red algae and used among other things to grow bacteria in a lab environment.

Blade: A broad, flattened leaflike structure on algae.

Chlorophyta: Phylum of the green algaes.

Extracellular digestion: Food digested outside the cell by passing enzymes through the cell membrane.

Heteromorphic: Where sporophyte and gametophyte appear different.

Hold fast: The part of an algae that grips the substrate and holds the plant fast.

Isomorphic: Where sporophyte and gametophyte appear alike.

Lichen: A symbiotic relationship between an algae and a fungus. Creates a new entity.

Macrocystis: One of the largest algaes. Can be over 100 feet.

Ozone: Oxygen in the O_3 state. Filters out harmful ultraviolet rays.

Phaeophyta: The phylum of the brown algaes.

Rhodophyta: The phylum of the red algaes.

Sargassum: A major type of algae found in warmer waters.

Silica: A compound similar to glass. Found in diatoms.

Stipes: The portion of a kelp plant between the blade and the base.

Thallus: A plant having no roots, stems, or leaves. Algae.

Zostra: One of the more widespread types of the sea grasses, commonly called "eel grass."

There are several ways to distinguish a plant from an animal, none of which work all of the time. If the cell has a wall containing cellulose, it is for the most part considered to be a plant. If the cell contains chlorophyll it is chiefly considered to be a plant. If it does not move when you touch it, it may be a plant. If it creates its own food through photo- or chemosynthesis it is generally a plant. We could go on but I think you have the idea by now. Plants, animals, and any other living thing, although different from each other, are all similar enough that there are no hard or fast rules to delineate them.

Plants were one of the first living things on earth, and through the process of photosynthesis, plants created an oxygen environment for animals to develop. This free oxygen also combined from the O_2 state, which is the state most oxygen used by animals is in, to the O_3, or **ozone** state, that filters out ultraviolet rays from the sun as they pass through our atmosphere. These ultraviolet rays cause cellular damage to most living organisms if not filtered out. A present-day concern of scientists is the diminishing ozone layer over the polar regions of the earth. Is air pollution breaking down the ozone? We do not know for sure, but we need to find out as soon as possible, for disastrous results could ensue.

KINGDOM MONERA

Taxonomists have placed the true bacteria and the blue-green algae together in this group because they are similar in having no nucleus and only one chromosome. They are composed of one cell of the prokaryotic type. (*Pro* means before; *Karyotic* means nucleus.) This is the first type of life on earth for which we have evidence. These life forms have been functional for about 4.5 billion years; we know this through fossil records. We will not go into any details about this group except to say they are an important part of the food chain of many benthic organisms. They create available food sometimes where little else exists.

KINGDOM PROTISTA

All the organisms except the Monera discussed above have eukaryotic-type cellular structures. Thus, that feature cannot be used to distinguish them. However, the fact that the organism consists of a single cell can be used. In this kingdom are included the **protozoa** (one-celled animals), the **dinoflagellates,** and the one-celled algae. We will discuss the protozoa and the dinoflagellates in a later chapter. It is the algae we are interested in here.

The one-celled algae consist of the *golden brown,* the *yellow green,* and the *diatoms.* These three comprise the phylum Chrysophyta. Also in this kingdom are the dinoflagellates of the phylum Pyrrophyta. The dinoflagellates are also classified by many as protozoa, and we will discuss them later in the text. Members of the phylum Chrysophyta are extremely important members of the plankton grouping. They almost single-handedly comprise the base of the planktonic food chain.

Diatoms

There are about 12,000 living species and 50,000 or more extinct species of diatoms. Off Beaufort, North Carolina, 369 species were identified in just two rather small mud samples. This is evidence of just how prolific diatoms really are. Diatoms have shells composed of polymerized, opaline silica. The shell of most forms consists of two parts, which allow for expansion. (See Figure 7–6.) Reproduction is both sexual and asexual and some benthic species are not photosynthetic, but rather absorb their food. Some species have lost their shells altogether and live symbiotically in species of foraminifera protozoans, where the diatom helps to produce food for its partner. Most diatoms, however, are pelagic planktonic forms and are the major base of the marine energy cycle.

Golden-Brown Algae

Golden-brown algae are prolific as nannoplankton. Because they are so small they were not thought to be of much importance until recently when better methods were developed to collect nannoplankton. With these newer collecting methods, the golden-brown algae were found to be a major food source in the marine planktonic community.

Yellow-Green Algae

This is a fairly small group of about 450 species, and they are predominantly freshwater organisms. One species, *Vaucheria,* commonly called "water felt" because of its physical appearance, grows on mud flats in tidal areas. It survives exposure to air quite well.

KINGDOM FUNGI

Fungi are heterotrophic plants that secrete enzymes through the plasma membrane and digest food outside the cell (extracellular digestion), and then absorb the digested material back into the cytoplasm through the plasma membrane. They have no means of locomotion and multiply by producing spores. Only about 500 species are found in salt water. They are all very small, 1 to 3 millimeters, and they play an important part along with bacteria in the decomposing of dead material in the environment. When a fungus combines with an algae the new composite organism is called a **lichen.** This is a *mutualistic* joining of two organisms to form a division of labor. Algae produce food and the fungus creates a protective covering that prevents the algae from drying out when exposed to air. This allows the pair to live intertidally where there is less competition.

KINGDOM PLANTAE

Plantae are multicellular plants consisting of two major subkingdoms, the **Thallophyta,** of which we are most interested because these plants compose the "seaweeds" of the world; and the subkingdom **Tracheophyta,** which has a few marine representatives but are mostly land forms.

Subkingdom Thallophyta

Cholorphyta (green algae), **Phaeophyta** (brown algae), and **Rhodophyta** (red algae), are all main participants in most all marine shallow environments. They are photosynthetic and are generally found near shore or on shallow shoals.

Chlorophyta: Green Algae

Over 7,500 species of green algae are known. They range in size from microscopic to over 25 feet long, and they are found in fresh and salt water as well as on land.

The genus *Codium* is common both on the West Coast (*C. fragile*) and the East Coast (*C. dichotomum*) of the United States. Because of its appearance, it is commonly called "sponge weed." *Ulva* is another common genus and is called "sea lettuce." *Ulva* can be up to several feet long, and half that width but only two cells thick. This thin **thallus** makes for a good microscope subject. The activity within the live cell can be easily observed. The life cycles of some of these algae are interesting because they illustrate **alternations of generations.** The organism moves from a diploid stage, which produces zoospores, that grow into a haploid state, which in turn produce gametes that join together to form a zygote that grows a diploid plant to start the cycle over. *Ulva* has such a life cycle. (See Figure 13–2.)

Phaeophyta: Brown Algae

Almost entirely marine, the approximately 1,500 species of this group are the most conspicuous plants of the intertidal regions of the cool water oceans of the world. Even in warm waters the genus **Sargassum** does well and is common. One of the largest kelps is **Macrocystis.** It has been known to grow to several hundred feet in length. It attaches to the bottom with a **holdfast,** has a stalk called a **stipe,** and a leaflike part called a **blade.** Most multiply through alternation of generations. Some of the common genera are *Nereocystis, Fucus, Laminaria, Ectocarpus,* and those mentioned above.

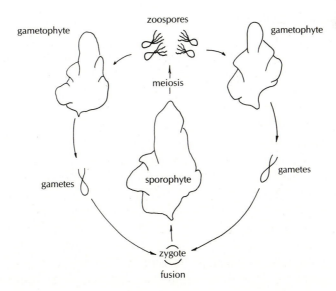

Figure 13–2 *Ulva* life cycle showing alternation of generations.

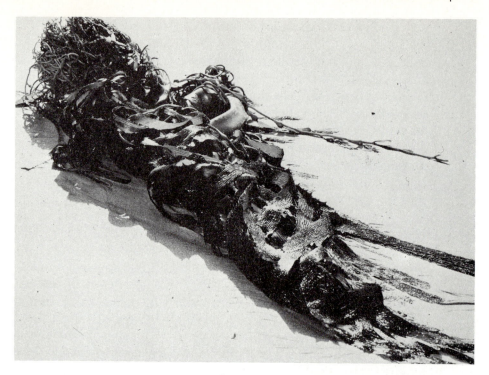

Figure 13–3   This entire kelp plant (Macrocystis) has been torn from the bottom by a storm and washed up on the beach. Many small organisms will use its decaying body as a home and for food. Nothing is wasted.

Rhodophyta: Red Algae

The red algae are better adapted to warm water than most other types. A few species, about a hundred or so, are found in fresh water but over 4,000 are found in salt water. There are more marine red algae than all the other marine algae combined. They are complex in structure and in life cycles. One of the main characteristics of the group is that there are no centrioles in the cells. Red algae have a reddish pigment that gives them their name, and the pigment is sensitive to that part of the light spectrum that penetrates deeper into the water than do other spectra. This allows the red algae to live in deeper water than all of the other algae. Some live as deep as 150 meters. Some species can deposit calcium carbonate in their cell walls and contribute to the building of the coral reefs. These are called *coralline* algae.

Most of these forms have alternations of generations with both **sporophyte** and **gametophyte** forms appearing similar. This is called **isomorphic.** If these forms appear different, they are said to be **heteromorphic.** All red

Figure 13-4 Some of the larger kelps have strong hold-fasts that attach them firmly to the substrata even during heavy wave action.

algae store food within the cell in the form of "floridean starch." This is an insoluble carbohydrate that is characteristic of the red algae.

Economics and Algae

Algae have been gathered and cultivated for centuries. Many thousands of people work in jobs just farming one red algae—*Porphyra*. The product is called *nori* and is a common feature of the diet of many Northern Pacific rim nations. The giant kelp *Macrocystis* is harvested to produce **alginates,** which are used as thickening agents in hundreds of products from food to paper. The substance **agar,** made from red algae, is used to make the capsules that contain drugs and vitamins, to grow bacteria cultures in the lab, to make jellies "gell," and for hundreds of other such uses. **Carrageenana,** a substance similar to agar, can be produced from the common red algae *Eucheuma isiforme* found in Florida. Most cultivated algae are at present in the Orient, but the Western world is slowly learning the methodology and the economic potential of this marine life.

SUBKINGDOM TRACHEOPHYTA

There are very few *seed plants* that can be called marine plants; however, the genus **Zostra** is one of the most widespread. This "eel grass" grows mainly where it has calmer waters. It acts as a giant filter to settle out silt as the rivers run into the estuaries. *Zostra* flowers in summer then dies. The decaying plants help to fertilize the water. Where pollution has killed the *Zostra* community in an area, severe changes have taken place in the environment. Everything from the loss of a fishery in the area to the silting in of a harbor has occurred.

Other flowering plants, which, like *Zostra,* are "sea grasses," include *Halophila,* found mainly in the Florida Keys; *Phyllospadix,* or "surf grass," which is found the entire length of the Pacific coast; *Thalassia,* or "turtle grass," of the extensive Florida shallows; *Syringodium,* or "manatee grass," of the southeast coast from Florida to Louisiana; and *Diplanthera* which is found from the southern part of Florida south.

To these sea grasses we need only add the mangroves to have all the flowering marine plants that occur in our part of the Northern Hemisphere. Mangroves are indicator organisms of a tropical environment. The southern coast of Florida is one of the finest examples of a true mangrove swamp. It is hundreds of square miles in area. In the tangle of roots there live organisms that are found nowhere else. Thirty or more species grow around the world. Some can live in deeper water than others, which helps to cut out some of the competition for space. Because roots need aeration, the air roots of the mangrove help it get air to the roots on low tide. Some of them grow up and out of the mud to reach the air. These are highly modified plants and have adapted to the marine environment totally.

REVIEW QUESTIONS

1. Why is it difficult to distinguish plant life from animal life?
2. What two kingdoms are composed of one-celled organisms, and how do they differ?
3. What are the changes in science that brought the golden-brown algae to the attention of scientists?
4. What is highly significant about the answer to question 3 above?
5. What group of algae contains: the most species? the biggest plants?
6. What is the significance of the "sea grasses"?

Chapter Fourteen

PROTOZOA

DEFINITION OF TERMS USED IN CHAPTER 14

Encyst: The ability of an organism to form a protective layer around itself to survive a harsh environmental change.

Globigerina-ooze: One of the main types of sediments found on the ocean bottom, composed of the shells of organisms in the genus of Globigerina. These deposits are high in calcium.

Lorica: The shell-like outer structure of the organisms of the group Tintinnids.

Pseudopodia: An extension of the body of the Rhizopods, often called a "false foot."

Radiolarian-ooze: Sediments found in the deeper ocean (below 3,600 meters), and composed of the silica shells of radiolarians.

Red tide: A condition caused by an extremely dense population of planktonic organisms. Most commonly caused by Dinoflagellata.

Zooxanthellae: Algae often found living within another organism, generally in a symbiotic relationship.

The Proto-(first) zoa (life) are microscopic animals found in all areas where it is possible for life to exist. They are also widespread, with single species being found all around the world. Many of them have the ability to **encyst,** or create a shell-like covering over their bodies, so they can live through dry spells that would kill most other living creatures. Their main characteristic is that they are composed of only one cell. These one-celled creatures are divided into five subgroups or classes. The first group is easy to distinguish because they are all parasitic and have no means of moving themselves. This group is the class **Sporozoa.** The Sporozoa are thought to be the most widespread of all animal parasites. They cause diseases such as malaria in humans, but are less significant in the sea although they do parasitize marine animals as well.

The second group is the class **Suctoria.** This small group is characterized by the fact they are attached through a stalk to the substrate and feed by using tentacles that actually suck the protoplasm, or life fluid, out of other protozoans that they catch. They are highly interesting little creatures because they eat animals many times their size, but they are of little importance to the study of marine life. The three remaining classes are worthy of a more extensive review: the Ciliata, Rhizopods, and Flaggellata.

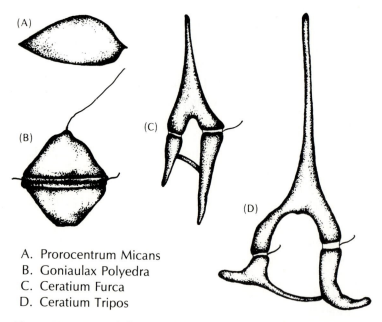

(A)

(B)

(C)

(D)

A. Prorocentrum Micans
B. Goniaulax Polyedra
C. Ceratium Furca
D. Ceratium Tripos

Figure 14-1   Typical shapes of West Coast "Red Tide" dinoflagellates.

THE CLASS CILIATA

The class Ciliata is characterized by having hairlike cilia throughout their life. These cilia are used to move them through the water, to catch food, or sometimes just to cause a circulation current for respiration.

One family of ciliates, the Tintinnids, is nearly all marine. These have a characteristic chitinous or pseudochitinous case that they secrete around themselves. The cases, called *lorica,* are different in each species and are used as the main identification feature. The lorica may not be protective in nature because some species which secrete no lorica seem to be as successful as those species that have them. This group bears mention because they are very numerous and normally planktonic. Commonly found in plankton samples, their number of species is probably in the thousands. One interesting sidelight is that some of them act as a host for a parasitic dinoflagellate. This parasite, when present, appears like a large, extra nucleus; but upon close inspection, it is seen to be a cell itself, with its own nucleus living inside the cytoplasm of the tintinnid.

THE CLASS SARCODINA

The class Sarcodina is characterized by movement accomplished by extending part of the body and then contracting it. The particular sarcodinids of interest to us are the orders **Foraminifera** and **Radiolaria.**

The Foraminifera ("hole bearers") have shells similar to snail shells,

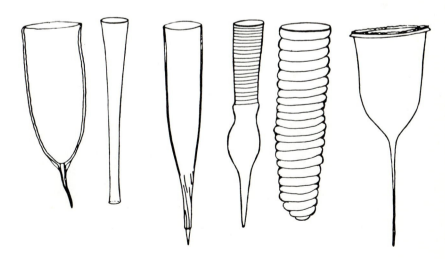

Figure 14–2 Atypical ciliata.

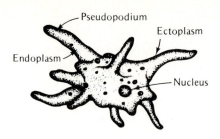

Figure 14–3 The Sarcodinids can change shape. Here is one showing many pseudopodia. In the orders Foraminifera and Radiolaria the shapes are more fixed.

composed of calcium and in some cases becoming porcelaneous as the organism matures. The shells have holes through which the protoplasm extends itself by flowing out and then back. This action, called *streaming,* creates **pseudopodia** (false feet). The pseudopodia are constantly moving, and a close inspection of them reveals small granules moving through the protoplasm. The terms *sol & gel* are used to distinguish the more solid outside portion (plasmagel). The streaming protoplasm is the main characteristic of the class. Some **forams** (short for Foraminifera) have shells of more than one chamber. As they grow, they add more chambers. Small holes are left between the chambers for the plasmasol to flow through and keep the organism from complete separation. These holes, called *stolon canals,* are smaller than the main openings used to form pseudopodia, which are located in the last chamber. Forams are larger than many other protozoans, with shells generally approaching 1 millimeter; some are even larger. Because of their relatively larger size, they are able to feed on smaller protozoans as well as on diatoms. A few types have a symbiotic relationship with algae called zooxanthellae, which live inside the cytoplasm and help to cre-

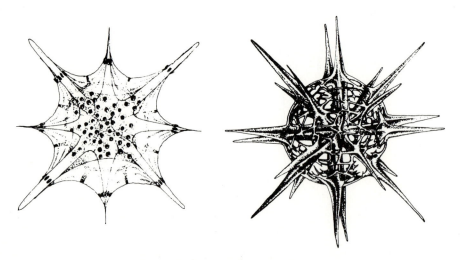

Figure 14–4 Two typical Radiolaria.

Figure 14–5 A characteristic Coccolithophore.

ate oxygen and use up waste produced by the foram. Over 15,000 species of forams have been classified, many of which are extinct and known to science only as fossils. These fossil forms are of great importance to the submarine geologist, who uses them as indicators of geological age when looking for oil-bearing strata. The main forams of interest to the marine biology student are species of the genus **Globigerina.** *Globigerina* has planktonic forms that are extremely abundant in the holoplankton. The shells of this genus have been deposited in the sediments of the ocean floor in such large amounts that they are the main component of large parts of the sea floor. These portions of the sediments are said to be composed of **Globigerina-ooze.** As the depth increases, the shells tend to dissolve; consequently, Globigerina-ooze is not found in the deepest parts of the sea, but it is found in most shallow to moderate depths. Foraminifera are also used to determine the temperature of the ocean during various ages in the past. The colder the temperature of the water in which they develop, the smaller and more compact the shells are and the smaller the holes for protoplasmic streaming. By studying the shells of forams taken from bottom samples and relating their depth in the sediment with time, the temperature of oceans can be reconstructed throughout our past history. This helps us to correlate glacial periods on earth and understand the general climatic patterns that have occurred in the past.

The second group of sarcodinids of special interest to us is the **radiolaria.** Radiolaria generally have a skeletal structure composed of silica, and they are round or spherical in shape. These forms are noted for their fine net type skeletons and are beautiful to look at. They are generally small, less than 2 millimeters, but are sometimes observed to cluster into masses 1 inch or so across. Their pseudopodia are similar in nature to those of the forams, as is their general feeding habit. Like some of the forams, they also are known to contain symbiotic zooxanthella at times. The main characteristic of this group is their siliceous skeleton. Unlike the calcareous skeletons of the forams the silica skeleton does not dissolve at depth. It is this phenomenon that accounts for most of the deep sections of the ocean having **radiolarian-ooze** as a base instead of Globigerina-ooze. This radiolarian-ooze generally occurs at depths greater than 3,600 meters and covers about

38,000,000 square kilometers (14,600,000 square miles) as compared to about 126,400,000 square kilometers (48,500,000 square miles) covered by Globigerina-ooze. An interesting feature of one of the radiolarians, the genus *Acantharia*, is that its skeleton is not silica, but strontium sulfate. Strontium, as an element, is barely detectable in ocean waters, but this organism is able to extract and concentrate it to form a skeleton. Many marine organisms have the ability to extract and concentrate the elements they require from their environment. Perhaps this will someday become important to humans when they learn how to extract these elements from the organisms.

THE CLASS FLAGELLATA

The class Flagellata (Mastigophora) is interesting for several reasons, particularly their basic classification. The zoologist considers them animals because they move about and show animal-like activity such as ingesting other animals for food. The botanist, however, considers them plants because they contain pigments that allow them to produce food from light, as do plants, and because they also secrete a plant-type material, called *cellulose*, as a protective covering. Instead of placing this group under the protozoa, it could be placed under the phylum Phyrophyta in the class Dinophyceae. This classification would designate it as an algae. It is placed here under protozoa more for convenience than for any other reason. Some instructors refer to them simply as plantamals. We are concerned here with their effect on the environment, not their proper taxonomic placement.

Of the several groups of flagellates, the most important marine group is the Dinoflagellata, which are characterized by two whiplike flagella. These small creatures make up a significant portion of the plankton and are

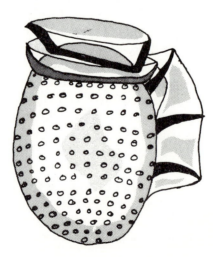

Figure 14-6 A highly magnified planktonic dinoflagellate, *dinophysis recurva*, found in the Mediterranean, the Atlantic, and the Coral Sea.

a major part of the basic food chain or energy transfer system. A few types produce an alkaloid poison, which is concentrated by some shellfish, such as the mussels, and is dangerous to humans who eat the mussel tissue. This is the reason for a quarantine on mussels in some areas during the part of the year when the dinoflagellates are abundant. During a particularly heavy bloom for which some species (*Goniaulax polyhedra* and *Gymnodinium breve* are well-known), the water turns a reddish color due to a red pigment in the organisms and a heavy concentration of them. This condition is commonly called **red tide.** Such a concentration of these Dinoflagellata and their related species not only secretes a toxin into the water, but also can use up most of the oxygen in the water and kill the fish in the area from a combination of these factors.

REVIEW QUESTIONS

1. What are the five classes of Protozoa?
2. What is the function of the pseudopodia?
3. Why are forams of interest to the geologist?
4. What organism is used to tell the temperature of ancient seas?
5. Why are dinoflagellata important to humans?

Figure 15–1 The name Porifera means "pore bearer." In this close-up of a sponge, the reason for the name becomes obvious.

Chapter Fifteen

THE PORIFERA

DEFINITION OF TERMS USED IN CHAPTER 15

Amoebocyte: A cell whose function is to carry food, waste, reproductive products, etc., within the sponge tissue from one place to another by migrating through the mesenchyme.

Choanocytes: Also called *collar cells.* These cells line the passages where water enters a sponge; they circulate water and ingest food.

Epidermal cells: The outside cells that give protection to the sponge like a "skin."

Hermaphroditic: Having both male and female gonads.

Mesenchyme: A nonliving gelatinous substance found between the layers of cells which adds thickness and form.

Osculum: The larger holes that can be seen in sponges. They are excurrent water openings, where water leaves a living sponge.

Pore cell: A constrictive cell used to close the opening in the sponge and regulate water flow.

Spicules: Smallest parts of the supportive structure or skeleton
of the sponges.

Sponges (Porifera), although not very dynamic, are still very interesting
creatures. The term *Porifera* means "**pore-bearers.**" This group is characterized by having many pores and canals through which water is drawn by a
special type of cell, called the *collar cell,* which in many ways resembles a
flagellate. These collar cells are more properly called **choanocytes,** after a
subgroup of the flagellata, Choanoflagellata, which some believe to be ancestral to the sponges. Sponges were first recognized as animals by Aristotle,
but through the years following most people considered them plants. Not
until the use of the microscope was the true nature of the sponge revealed.
Whatever large proportion a sponge may reach, even 5 to 6 feet across, it
is totally tunneled by small canals through which the choanocytes keep the
water moving by the whiplike action of their flagella. The canals are lined
with choanocytes, and these cells capture and digest the microplankton
brought them by the circulation of the water through the canals. The water
enters the sponge from the outside and is moved inward to a larger cavity,
where it leaves the sponge through a larger hole called the **osculum.** It is
the osculum that we see when we look at the larger holes in the sponge. It
is interesting to place a small amount of dye in the water next to a sponge
and watch it disappear, only to reappear through the osculum. This method
is commonly used to study circulation through the sponges.

Because the sponges vary so much in shape, they are one of the more
difficult groups to identify as to species. Finding different shapes and colors
within the same species is not uncommon. These differences are caused by
the varying environmental factors affecting each individual and the individual's specific growth response to those factors. One characteristic, however,
can be used for identification with reliability. Most sponges contain a skeletal
structure composed of many thousands of tiny parts called **spicules.** By their
characteristic shapes, these spicules can be used to separate and identify
most of the sponges. According to the spicules, the sponges can be divided
into three major groups or classes. The types in the first class, Calcarea,
have spicules composed of **calcium carbonate,** are marine in nature, and
generally under 18 centimeters long. The second class, Hexactinellida, has
spicules composed of **siliceous material.** These sponges can reach a size to
1 meter and are normally deep water marine organisms. Some fairly shallow
species occur in the Antarctic and Japan, but most of them are found below
the 300-meter level. The third class, Demospongiae, normally has **spongin
fibers** along with or instead of spicules. Spongin is a sulphur-containing protein that is resistant to most environmental factors, including digestive en-

Figure 15–2   The sponge maintains its shape due to the spicules that serve as the skeletal structure.

zymes. This is the group that has historically been harvested as bath sponges.

Sponges seem to be an entity unto themselves. They are a side group on the evolutionary progression and, although very successful, give no evidence that they are ancestral to any other animal group.

To the observer, the members of the Porifera are aesthetically interest-

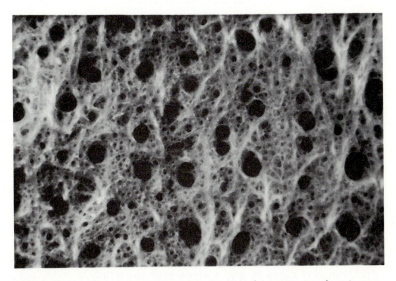

Figure 15–3   One can see the great porosity of sponges on close inspection. The water is taken in through small pores on the outside, filtered through the organism, expelled through the large openings.

ing because of their widely varied forms and sometimes striking colors. They are not something that would warrant a long period of observation, because they show no obvious movement, even when physically handled. They are capable of closing some of the pores in their bodies, but this is done very slowly and is quite often too small a movement to be noticed. This lack of visible movement is what caused the confusion as to their rightful placement in the animal kingdom.

The sponges have several types of cells; each has its specific function and is a good example of what is termed "**division of labor.**" The choano-cytes circulate water and gather food. The pore cell is the cell at the entrance to each pore in the body of the sponge. These pores are microscopic, not the large holes you see when you look at a sponge. Each pore cell is shaped in a ring or is in conjunction with other pore cells so that they can contract and close off that particular pore. The outer covering of the sponge is generally composed of a special group of cells called the **epidermis.** These covering cells fit together in much the same way as tiles fit to cover a floor. Between the epidermal cells and the choanocytes is a nonliving material called **mesenchyme,** a gelatinous substance that acts as a filler between the outer and inner layers of cells. In the mesenchyme, and able to move through it, is another type of cell called the **amoebocyte.** The amoebocyte acts as a carrier. It carries food from the choanocytes, where it is gathered, to the epidermis and pore cells, and also carries waste products back to the choanocytes to be cast out into the water. Although the Porifera have no circulatory system, the amoebocytes perform most of the same functions. This division of labor becomes more complex in higher animals and is used in part to distinguish the more ancient forms of life from the modern ones.

Sponges have adapted well in most salt water environments. There are sponges in Alaska that are so mud-filled that biologists wonder how they can circulate enough water to survive. Off the coast of France a sponge grows in the shape of a sea fan, and back in Alaska there is one that resembles a pineapple. In Southern California, one type of sponge grows in long pads that cover the rock. Some of these pads are over 3 meters long, 1 meter wide, and a 0.3 meter thick.

Because the spicules are like microscopic thorns, very few animals eat sponges. Their chief predator is the nudibranch, a type of mollusca. Because sponges lack predators, certain crabs have found it beneficial to place small pieces of sponge on their backs. As the sponge grows, the crab's predators leave it alone. For the hermit crab, the sponge often gets so large that the crab cannot carry it any longer and must leave it for a new shell. Many animals use the sponge as a home. Certain annelid worms, shrimp, brittle stars, and many others can be found living in the cavities of a large sponge. For the most part, the sponge is a passive landlord deriving no particular benefits from the majority of its residents.

Most sponges contain both male and female sex organs; that is, they are **hermaphroditic.** Many animals have this trait, which makes every member of the species a potential producer to maintain the population. Normally, however, they do not fertilize themselves. The eggs and sperm develop at different times, and the sperm are released into the water to be filtered into the pores of other sponges. They are then picked up by a collar cell and carried by an amoebocyte to a mature egg. The resulting larvae go out with the excurrent water, joint the plankton for a short time, then settle to the bottom to develop into a parent type. Any piece of sponge is capable of growing into a new sponge, so there is also **asexual** reproduction through *budding.* Budding generally increases the size of a sponge, because the new bud will stay attached and grow from the side of the parent sponge.

REVIEW QUESTIONS

1. What are the two main functions of the choanocytes?
2. Why are the sponges difficult to identify as species?
3. What is the difference in spicules in the three classes of Porifera?
4. What are the main types of cells found in sponges?
5. How does the sponge keep from fertilizing itself?

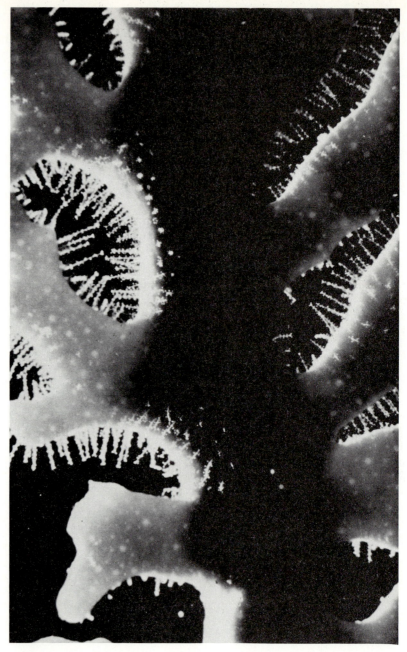

Figure 16–1 Sea forms are often beautiful and many have vivid colors.

Chapter Sixteen

CNIDARIA

Anthozoa: Corals and anemones belong to this class.
Gorgonians: A type of colonial anthozoan that forms the sea fans.
Hydrozoa: The various hydroid forms belong to this class.
Mesoglea: A clear, noncellular substance found between the epidermis and the gastrodermis in some coelenterates.
Nematocysts: Stinging capsules used by most Cnidoria for protection and securing food.
Polymorphism: Ability for one species to occur in several different shapes or forms during its life cycle.
Scyphozoa: Jellyfish belong to this class.

The **Cnidaria** are highly visible to the beach walker, diver, angler, or sailor alike. Although they are no more important in the general biological plan of the marine environment than other organisms, they are generally studied to a greater extent because of their accessibility to most interested people. Members of the Cnidaria are found in cold waters of the world as

well as in the tropics; many are large and sedentary so they are easily observed. Some of them have wrecked ships, others have killed people, and
still others are known mainly for their sting. They build islands, are common
subjects for underwater photographers, and even lend themselves to creating fine and expensive jewelry. Unlike the Porifera, which are of little interest to the general public, the Cnidaria are known in one form or another to
people everywhere.

The phylum Cnidaria is generally divided into three classes, each of
which has a wide distribution on its own. These three groups include the
sea anemones and the corals in the class Anthozoa, the jellyfish in the class
Scyphozoa, and the hydroids in the class Hydrozoa. They share the same
general characteristics: a digestive cavity with only one opening, radial symmetry, two layers of cells with a jellylike substance called **mesoglea** in-
between the layers, and stinging capsules called **nematocysts.** Also, they
lack special organs for respiration and excretion and have no blood. These
traits group them as Cnidaria, while other specific traits separate them into
their individual classes. The general body shape of each class is one of the
main separating characteristics.

With the exception of the common fresh waterpolyp *Hydra,* and a few
less common forms, all Cnidaria are marine. Many of the groups are highly
colonial. These colonial forms join together and in some cases have a common digestive tube. They create tree or fan-shaped colonies up to 3 meters
across consisting of thousands of small polyps (hydroids), which, because of
their small size, appear to be featherlike. Colonies of coral, which form a
hard calcium case around each polyp, create the coral reefs of the world.

A basic characteristic of the phylum is the presence of a stinging cap-

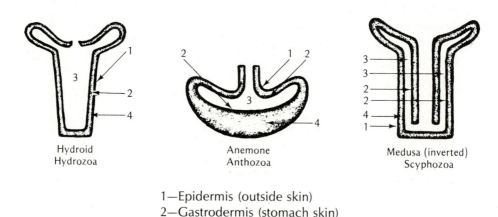

Hydroid   Anemone   Medusa (inverted)
Hydrozoa   Anthozoa   Scyphozoa

1—Epidermis (outside skin)
2—Gastrodermis (stomach skin)
3—Enteron (intestine)
4—Mesoglea (middle-glue)

Figure 16–2 Body shape by class of Cnidaria.

Figure 16-3 The small organisms which make up each of the different colonial forms are easily recognized by a student of Cnidaria. Here we see examples of several different types.

sule called a *nematocyst*. These are found on almost all parts of the body, but the greatest numbers are on the tentacles. Often grouped together to form a "battery," they are responsible for catching food as well as protecting the organism. Some of the nematocysts are so strong they can cause severe skin irritation to humans. Among the worst types for humans to encounter is on the **Portuguese man-of-war,** *Physalia,* which has a very toxic sting that can require hospitalization if a person becomes entangled and receives many stings. Jellyfish often sting swimmers that swim into them, causing rashes and discomfort for a few hours. The nematocysts of different species are of varied shapes, sizes, and toxins. The toxin is a protein poison carried at the head end of the stinger. The cells are triggered by chemical as well as mechanical means and do not regenerate once fired. New cells are formed and migrate through the tissue to the area where they are needed for replacement. Many years ago G. E. MacGinitie demonstrated to the author a simple microscopic technique for watching the nematocysts perform. It involved smashing a small portion of a tentacle on a glass slide with a scalpel. The slide was dried and then flooded several times with pure alcohol and allowed to dry between each flooding. A coverslip was placed on the dry slide, and with fairly high power, a group of nematocysts were located. A drop of methylene blue solution was flooded across the slide under the coverslip to stain the nematocysts and make them easily visible. The osmotic pressure of the solution being absorbed by the cells made them "shoot off." The threads that hold the victim to the Cnidaria were then visible and at the end of some of them a drop of toxin could be distinguished. This simple laboratory procedure can be performed by almost anyone in a school lab.

THE CLASS ANTHOZOA

The corals, sea anemones, and sea pens are common forms worldwide, the corals being found in warm and temperate seas and the anemones in all seas. The term *anthor,* from the Greek, means "flower." The members of this group do resemble flowers. All of this class attach to the substrata, have hollow tentacles, and for the most part, are of moderate size. They feed on small animals that they immobilize with their nematocysts as do all Cnidaria.

The sea anemones vary in color; some are bright red, green, or orange, whereas others are snow white. Some large ones in British Columbia are 1.3 meters high and 0.5 meters in diameter; others are barely 1 centimeter across. Many intertidal forms catch and hold sand and small bits of shell against their bodies for protection. When the tide is out and they are exposed and totally retracted, they appear to be a layer of small bits of shell covering the rock. The deeper forms generally do not use this type of covering.

The mouth opens into a gullet and not directly into the enteron; this is a main characteristic of the class. Food is passed into the mouth by means of the tentacles and the cilia on the tentacles, which push toward the end

Figure 16–4 The sea anemone is closed up tight to protect it from drying out while the tide is out. Tiny shells and sand particles stuck to its body give it protection and make it harder to see.

Figure 16–5 This close-up photograph of the tentacles of a sea anemone shows the number and general shape of the tentacles.

of the tentacle. When a nonnutrient falls on the tentacle, it is carried to the end of the tentacle and drops off. When a nutrient is dropped onto the tentacle, however, the tentacle is curled in and the substance is carried to the end of it but falls into the mouth. Although the cilia along the gullet normally beat in an outward direction to keep water flowing for respiration, when food is present, the cilia reverse direction and take the food down the gullet to the enteron. When the food has entered the gastric area, the cilia resume beating in an outward direction to maintain respiration. Most anthozoa reproduce both sexually and asexually. They develop a ciliated larvae that swims in the plankton for a short time before settling down to a solid substrate and developing into an adult. Anenomes are solitary as well as colonial. Although most live on rock, some species burrow into the sand or mud. Some of these burrows are over 15 inches deep, and the ability of the animal to contract 12 to 15 centimeters beneath the surface when disturbed makes it difficult to dig out without injury. Intertidal members of this group are quite resistant to unstable conditions and live well in aquariums. One specimen of record has lived as long as 66 years in an aquarium.

The sea pen and sea pansy also belong to this class. Both of these colonial polyps form a regular-shaped colony. Found in mud or sand, they have a stalk extended into the substrate to anchor them. The sea pansy is a flattened colony about 5 or 7 centimeters in diameter and looks like a leaf with a short 5-centimeter stem. The sea pen may extend over 5 meters into the substrate and sticks up above the sand or mud from several inches to several feet. When the polyps on the sea pen are extended, it looks much like a large bird feather sticking in the bottom. Some sea pens that do not extend so deeply into the bottom can uproot themselves and change locations. These are the main forms seen by divers or collected by dredge.

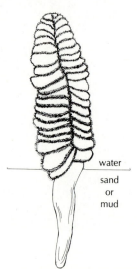

water

sand
or
mud

Figure 16–6 The sea pen grows in burrows with its feeding polyps above the surface of the bottom to collect food.

Figure 16-7 The hard corals build a hard shell around their bodies and, in doing so, form reefs.

Figure 16-8 The sea fan is one of the more striking colonial Cnidaria forms.

The corals are also anthozoans. How they build reefs was discussed in the section on equatorial areas. Here we will discuss the general life style and form of various members of the group.

Generally referred to as gorgonian corals, this group belongs to the order Gorgonacea. These are colonial forms that grow from a central or axial skeleton. They appear plantlike, as the polyps are small and many thousands generally make up a single sea fan. One type, *Corallium,* is a red coral, and the skeleton is used to make expensive jewelry. Divers often take home sea fans because they are of such delicate beauty in the sea, only to find they smell very bad (the rotting polyps) when dried. They are generally discarded and the environment has lost an entire colony of organisms. These corals form a skeleton; this is what allows them to assume their various forms. Different species are easily recognizable from their different shapes. In the Caribbean the gorgonian and soft corals are so prolific and varied as to type that the bottom appears to be a garden of long fingers and sea fans waving in the current.

Soft corals are colonial polyps without axial skeletons. The polyps extend from a fleshy base that protrudes on a stalk, and the colony waves like giant, boneless fingers with the current.

Hard Corals or Stony Corals

The great reefs of the world consist of the skeletons of the hard corals as a base. Although a coral reef has many components, its basic mass is the skeletons of the hard coral. The living animal is very much like an anenome. The main visual difference is the hard calcium cuplike house the coral organism builds to live in. The majority of reef-building corals need water temperature of at least 20°C (68°F) to flourish, a requirement that limits them to the warm water of the equatorial belt. This belt lies south of mid-Florida on the East Coast and of mid-Baja California on the West Coast of the United States (approximately 28° north latitude) and north of the Chile-Bolivia border on the western coast and of Rio de Janeiro on the eastern coast of South America (approximately 23° south latitude). Many hard corals occur in colder water, but they are not the reef builders of the warm seas. Some of the solitary corals occur in California, and others in the deep ocean at depths below 6,000 meters. The largest reef in the world is the Great Barrier Reef off the east coast of Australia. It is 1,900 kilometers long and has formed hundreds of small islands along its length. A thousand times larger than any other structure built by a living creature in the world, these reefs are constructed of coral growing on top of coral, smothering the ones below and then being smothered themselves by new ones growing on top of them. Most of the coral reefs we know today were constructed within the last 30,000 years. As the reef grows, it makes an excellent home for thousands of

Figure 16–9 This "Sailor-by-the-wind," Vellela, has air cells in its body and a sail-like portion which allows the wind to move it across the surface of the water. It often washes ashore and can be found on the beach.

other organisms. Fish and all other types of marine animals find microenvironments within the coral reef that suit their adaptations. Some coral is so dependent on the presence of an algae that grows on the reef that they cannot exist without it.

Although the reef building corals have few enemies, in recent years humans have destroyed large areas through the dredging of new harbors. The coral is smothered very quickly by silt settling out of the water upon it. When humans dredge a new harbor, they dump the silt out into the sea around it. Such dredging has killed large areas in Grand Cayman, Hawaii, and Mexico. At the present time, a large harbor is proposed for construction in a swamp in Panama. If this harbor is built, many miles of coral will be destroyed. One of the natural enemies of coral is the crown of thorns starfish in the Pacific. It has destroyed many reefs during the 1960s and 1970s. There has been an active starfish control program going on that has had varying results. Some claim it has helped; others claim it has not. It appears at this time that it has helped, but that the problem may be one of nature's many life cycles and that humans have just become aware of it.

THE CLASS HYDROZOA

Most marine forms are colonial. Some have strong nematocysts and can cause irritation and pain if the bare skin comes in contact with them. They are for the most part, very delicate, lacy, bushlike structures that attach to the bottom. The colonies are often mistaken for algae. Within a single colony can be several different kinds of polyps. Generally, there are feeding polyps that gather food; defensive polyps that do the stinging; and reproductive polyps that produce a different form called a *medusa* for sexual reproduction. The medusae, although totally different in appearance, are just a reproductive form of the same animal. This **polymorphism** (that is, different shapes and functions within a single species) shows a type of division of labor that reaches its peak in the bodies of the higher animals, such as the mammals.

Because of their frail construction, the hydroid colonies are generally found in water that is not very turbulent. Bays are an excellent habitat for them, as is the offshore water where there is little wave action. In Newport Bay, California, some solitary hydroids grow to 7.5 centimeters and resemble small transparent palm trees growing out of the mud. One group deserves special consideration. The siphonophores are colonial hydroids that are free swimming. The common one, well-known off the East Coast of the United States, is *Physalia,* the Portuguese man-of-war. It has very toxic defense polyps with powerful nematocysts that can send a swimmer to the hospital, as we mentioned earlier. On the West Coast the genus *Velella* is common but does not sting to any extent on contact. Both these forms float on the surface and are beautiful to observe being blown along by the wind as it hits their ''sail'' protruding above the water.

CLASS SCYPHOZOA: THE JELLYFISH

Everyone is familiar with the jellyfish. The opinion of the general public is that they are beautiful to look at, but dangerous to touch. In general, this belief is true. The siphonophore of the warm eastern coastal waters, the Portuguese man-of-war, is known to most laypeople as a jellyfish, even though it is not, as we now know, scientifically included in this group. Whereas the siphonophores are colonial, the jellyfish is generally a single organism. The form in which scyphozoans are normally seen is called a **medusa.** The medusa is bell- or umbrella-shaped and is free floating, with limited swimming ability. It achieves the swimming effect by pulsating its bell-shaped body and pushing water out in a jet-type action. This generally slow movement is used mainly to control depth. Its body is composed mostly of the clear mesoglea that has given them the name ''jelly.'' The term ''fish'' is unfortunate, but too ancient in use to be changed now. Members of this group range in size from around 2 centimeters to well over 2 meters in diameter and may have tentacles over 30 meters long hanging from the inside of the bell near the mouth. The largest specimens are generally seen in the northwest Atlantic Ocean, but the author observed one unidentified specimen in the mid-1950s off Catalina Island, California, that was 1.8 meters across, with tentacles of 12 meters. The tentacles, as in all Cnidaria, catch food and contain many nematocysts. The larger forms can sting through the human skin and cause discomfort. During storms or heavy surf conditions, jellyfish are often cast ashore. Beachcombers find them or parts of their broken bodies on the beach. Many people have discovered, much to their dismay, that the nematocysts do not cease to function when the animal dies. Picking up pieces of a dead jellyfish can be a surprisingly painful experience.

The sexes are separate, and sexual reproduction takes place between the male and female medusa. After the sperm swims through the water to fertilize the eggs in the female, a small ciliated larva swims in the plankton, then settles to the bottom. It grows slightly and then segments into medusa-shaped parts that invert and grow into an adult medusa. In a few species, such as *Pelagia,* the egg develops directly into a medusa form and is fixed to the bottom during its entire life. Most Scyphozoa, however, **alternate generations** of fixed and free-swimming forms.

The jellyfish have few enemies, the most visible one being the ocean sunfish (*Mola mola*). This predator, weighing as much as 1,000 pounds, drifts in the plankton as the largest planktonic form and lives mainly on jellyfish. It swims rather poorly and is found in great numbers in water with a high jellyfish population.

One of the most widely distributed jellyfish is **Aurelia.** It is nearly world-wide, and although it does poorly in warm water, it has been frozen in ice and still survived. Its ability to adapt to a large range of temperatures may well account for its wide range. The jellyfish, and the giant squid are thought to be the largest of the invertebrates. Although some other invertebrates may be longer, probably none weigh as much as a large North Atlantic *Cyanea* or giant squid.

One other Cnidaria of this group is also of interest because of its highly toxic nature. These are the sea wasps. They are small (2 or 3 centimeters), and have a bell shape more like a box, with delineated corners. They have one or two tentacles hanging from each corner of the bell and are so toxic that they have caused many deaths in the area of Australia, Borneo, and Malaya. They also occur in the Caribbean, but not in dense populations; they have not been a problem there. Hard to see because of their size and transparency, they are more likely to be a danger to scuba divers and swimmers than to anyone else.

REVIEW QUESTIONS

1. How do the small hydroid forms create large fans?
2. What is a nematocyst?
3. How do the gorgonians differ from the stony corals?
4. What are some of the differences between the Hydrozoa and the Scyphozoa?
5. Why is it unwise to pick up parts of a jellyfish one might find on the beach?

Figure 17–1 The "tube snail" is one of the mollusca that grows with its shell attached to a rock.

Chapter Seventeen

MOLLUSCA

DEFINITION OF TERMS USED IN CHAPTER 17

Bivalvia: Class name of the clams (formerly called Pelecypoda).

Byssal threads: Means by which the sea mussel attach to the substrate.

Cephalopoda: Class name of the squids and octopus.

Chromatophores: A cell on the skin of an animal that can expand and contract to change color.

Conus: A widespread genus with many species, some of which are dangerous to humans.

Gastropoda: Class name of the snails.

Littorina: Very common genus of gastropod often used as an indicator organism to locate tidal zones.

Loligo: A very common genus of squid.

Nudibranchs: A group of mollusks with no shell and their gills exposed.

Operculum: A hard plate attached to the foot of some Gastropods that enables them to close the opening of their shell.

Periostracum: A thin, tough outer layer on most bivalvia shells.

Piddocks: A boring clam that causes damage to submerged marine structures.

Polyplacophora: Class name of the chitons (formerly called amphineura).

Pteropoda: A swimming gastropod.

Radula: A hard band with minute teeth used by most mollusks to scrape food off rocks. Not found in bivalvia.

Scaphopoda: Class name of the tusk shells.

Teredo: Common shipworm causing great damage to any wood structure in the marine environment.

Tethys: Genus name for the common sea hare.

Umbo: The hinge area of a bivalvia.

Because of structure, the phylum Mollusca is not only well-known but widely collected. The Mollusca are the seashells that people pick up and admire when visiting the beaches. Because of their beauty, a great many books on shells have been written and beautiful photographs taken of them. Seashells are made into jewelry, lamps, ash trays, flower pots, dishes, and a great many other things. The rare ones sell for as much as $2,000, and the common ones cover the beaches. It is only human nature that we know so much about the shells and so much less about the animals that create them and live in them until they die. The molluscan shell is constructed of **calcium carbonate** secreted by a "gland" called the **mantle.** In 1943, Tracy I. Storer described the characteristics of the mollusca in his text on zoology. His description is duplicated below to prove a point. Read it carefully and realize as you read that the shell is really a very small part of the animal.

1. Symmetry bilateral (viscera and shell coiled in GASTROPODA and some CEPHALOPODA): 3 germ layers; no segmentation; epithelium 1-layered, mostly ciliated and with mucous glands.

2. Body usually short, enclosed in a thin dorsal mantle that secretes a shell of 1, 2, or 8 parts (shell in some, internal, reduced, or none): head region developed (except SCAPHOPODA, PELECYPODA): ventral muscular food variously modified for crawling, burrowing, or swimming.

3. Digestive tract complete, often U-shaped or coiled; mouth with a radula bearing transverse rows of minute chitinous teeth to rasp food (except PELECYPODA): anus opening in a mantle cavity; a large digestive gland ("liver") and often salivary glands.

4. Circulatory system includes a dorsal heart with 1 or 2 auricles and a

ventricle, usually in a pericardial cavity, an anterior aorta, and other vessels.

5. Respiration by 1 to many gills (ctenidia) or a "lung" in the mantle cavity, by the mantle, or by the epidermis.

6. Excretion by kidneys (nephridia), either 1 or 2 pairs or 1, connecting the pericardial cavity and veins; coelom reduced to cavities or nephridia, gonads, and pericardium.

7. Nervous system typically of 3 pairs of ganglia (cerebral above mouth, pedal in foot, visceral in body), joined by longitudinal and cross connectives and nerves; many with organs for touch, smell or taste, eyespots or complex eyes, and statocysts for equilibration.

8. Sexes usually separate (some hermaphroditic, a few protandric): gonads 2 or 1, with ducts; fertilization external or internal; mostly oviparous; egg cleavage determinate, unequal, and total (discoidal in CEPHALOPODA), a veliger (trochophore) larva, or parasitic stage (UNIONIDAE), or development direct (PULMONATA, CEPHALOPODA); no asexual reproduction.

The next time you look at a shell collection, try to think, not just how spectacular the shells are, but how unique the animals were that made and lived in them.

There are seven classes of living mollusks. Because five of them are commonly encountered, the names should be committed to memory, along with the characteristics of each group. They are Polyplacophora (Amphineura), the chitons; Gastropoda, the snails; Bivalvia (Pelecypoda), the clams; Cephalopoda, squid and octopus; and Scaphopoda, the tusk shells. The sixth class, Aplacophora, consists of small, worklike organisms that are benthic. The seventh class, Monoplacophora, is the smallest shelled animal. It was discovered in the early 1950s at a depth of around 3,500 meters, taken in a dredge by the ship *Galathea* off the Pacific coast of Mexico. Classes 6 and 7 will not be discussed here because of their rarity.

THE CLASS POLYPLACOPHORA

The chitons are among the most primitive of all the mollusca. They are easy to recognize as they have shells that fit together so they can bend. They are all marine and have overlapping **eight plates** that cover their back and enable them to adapt to the shape of the substrate. Able to withstand heavy surf impact, they are common tide pool organisms. Their gills are well hidden on either side of the foot under the shell. Using a **radula,** they com-

Figure 17–2   The chitons can be recognized by the series of plates that make up their shell.

monly graze on algal film. The radula is a mouth part found in all classes except the Bivalvia. It is attached to the floor of the mouth and can be extended to scrape algae off the rock. It is much like a file, having many rows of fine teeth. The chitons are nocturnal and crawl to feed only at night. They do not go far and generally return to the same resting place each day. The algal film on which they graze is composed of diatoms and other minute organisms that settle on all surfaces every day, a food source that is constantly being replenished. The largest chitons are 23 centimeters long and 12 centimeters wide, but most are only 2 to 4 centimeters long.

THE CLASS GASTROPODA

The gastropoda are a large and interesting group. Their name, meaning "stomach foot," comes from their body shape, with a **single shell** carried on the back and the large, broad foot on which it crawls across the rocks or digs into the sand or mud. The common forms are the snails, limpets, abalones, conchs, and whelks. Some forms less easily recognized as gastropods are slugs, nudibranchs, and the lesser known pteropoda and heteropoda that have adapted to a planktonic life. The gastropod has a characteristic spiral shell, which, with only a few exceptions (such as *Antiplanes* of the Pacific Northwest), turns to the right. The bodies of the animals show modification by a reduction in some of the organs to better fit the shell. A common example of organ reduction is the single kidney; in some only one gill is present. When we observe a limpet shell, we do not see a spiral although the shell was a spiral during the larval stages. As it settles to mature, the spirals are lost. The shell is almost nonexistent in some forms like the sea hare (*Tethys*) and is totally nonexistent in others known as nudibranchs.

In most areas the shell acts as protection for the animal. When disturbed, some animals, such as the limpets and abalone, pull the shell down tightly against the rock and hide beneath it. Others are able to retreat into their shell and close it off by use of an additional part, the **operculum.** Composed of calcium on some and softer horny materials on others, the operculum is a plate attached to the foot. As the animal withdraws into its shell, this plate is the last part to be pulled in and seals off the entrance. These opercula are interesting, as they show lines of growth and sometimes can be used to age the snail. The shells of most gastropods show lines of growth, but their age cannot always be determined by them. Some types grow slowly and evenly; others will add large extensions to their shells in a few days. Because these shell extensions come at irregular intervals, the age cannot be determined by growth lines on the shell. The shell of each species is different and characteristic of only that species. Most shell variations, if studied along with the habitat of the animal, can be recognized as specific adaptations to the animal's environment and life style. For example, low, flat shells are found on animals in areas of strong water movement where a high profile would create too much resistance. They are also found on animals that live in cracks or between rocks where the low shell enables them to crawl back into well-protected areas.

Some gastropods are more highly sought after by collectors because of their smooth, shiny shells. An example of this group is the cowries. They keep their shell glossy by running the mantle out over the shell to keep it clean. Collectors must be careful if they are to preserve the shell in its shiny state. The best method seems to be to place the cowrie into a disinfectant soap solution as soon as it is caught. This will prevent bacterial or enzyme action from deglossing the shell until it is cleaned and oiled for exhibit.

The majority of gastropods crawl forward on their food by gliding along as the land snail does. As always in every group, there are a few exceptions. The most exotic exception is the genus *Janthina*, which makes air bubbles by secreting mucus and trapping air near the surface and uses them for flotation. They are planktonic drifters. Some types use the modified foot for swimming; others use the operculum as a hook and reach out with it, then pull themselves along. Still others have a split foot. One type has a foot split in such a manner that it reaches out with the anterior portion, takes hold, then pulls the posterior section along. Another has a foot split down the center, from front to back. When it moves, one side goes forward then the

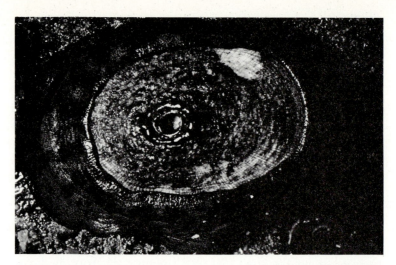

Figure 17–3 This "key hole" limpet has a single shell as do other limpets, but its shell does not cover its entire body and has a hole in the top. It is one of the largest forms of limpets.

other side. This creates a swaying action as it moves along. These various methods of movement and variation in shell form only reaffirm that with proper adaptation, an organism can live almost anywhere. The mud burrowers and some sand burrowers, like the olive shells, have an extended siphon that is a portion of the mantle modified to bring water to them even when they are beneath the surface. They need this siphon for respiration and because of it can inhabit a subsurface habitat where competition for space is not so strong. Several groups deserve specific mention, either because they are common or different.

The **pteropoda,** or wing-feet, are modified for swimming, and some of them are entirely planktonic. They often occur in large, dense populations. There are a number of species, some of which have no shell at all and are found in colder water, even where ice is present. The warmer-water species mostly have light calcium shells and contribute to the makeup of the various oozes.

The **opisthobranchiata** are sea slugs and normally lose their shell by the time they become adults. They are divided into two suborders: those with the gills covered by the mantle, the sea slugs; and those with the gills exposed to the surface, the nudibranchs. The sea slugs are the tectibranchs. They generally have a light shell or remnant of a shell. Some of these are called bubble shells. When the bubble shell animals move, their foot extends two or three times the shell size. Others have no obvious shell; the striped mollusk (*Navanax*) of the West Coast and the sea hare (*Tethys* and related types) are examples. The sea hares are **hermaphrodites** (both sexes)

Figure 17–4 A tube snail extends its foot to feed.

often seen in tide pools. The largest specimens seen by the author were taken off of Catalina Island in 1961 from 4.5 meters of water. The larger one weighed 35 pounds, the smaller one 33 pounds.

The **nudibranchs** are the gastropods that have their gills exposed. In reality, there are no true gills in the group. Respiration takes place through the epidermis on the gill-like structures. The color of these creatures is generally bright and magnificent. They are delicate things of beauty and grace when they are observed in life, but retain nothing of their beauty or grace when preserved. The only way to appreciate them is in an aquarium or by scuba diving into their environment. They commonly eat hydroids. It is

Figure 17–5 The eggs of the sea hare form a strong mass that resembles gelatinous spaghetti.

interesting to note that in some cases the nematocysts of the hydroid are not digested but migrate into the body tissue of the nudibranch and are used by it for protection. The nudibranchs are divided into two groups, easily determined by the gills that retract into an opening on one type (doridids) and do not retract on the other (eolid) type. If the gills are removed in the eolid type, respiration is carried on through the skin.

The **cone shells** need to be mentioned because they are widespread and possibly dangerous. The genus **Consus** consists of 600 or more species, most of which are collectors' items. The unusual modification of this genus is the radula, which has developed into hollow, sharp, often barbed teeth that are connected to a poison gland. In the larger cones, generally from the Western Pacific and Indian oceans, their sting and poison have caused human fatalities. They should be handled so as not to come in contact with the aperture of the shell. A sting, although rarely fatal, is often painful.

The wormlike mollusk and tube mollusk build tubes to inhabit instead of having a shell to carry around. The calcium tube is attached to rocks, shells, or some other solid substrate and sometimes grows in irregular shapes, depending on the room available for it to grow. This group has often been confused with the tube-building annelid worms. The main difference can be seen by looking closely at the tube. If it is shiny inside, it is a mollusk;

Figure 17-6   The abalone feeds by lifting the shell off the rock and "grazing" for algae growth on the rocks nearby. Abalone move around mainly at night, spending days in cracks and under rocks.

if it is dull, it is an annelid. In addition, the tube of the mollusk is three-layered, while the tube of the annelid is two-layered.

The **abalone** provides an important fishery along the Pacific coast. A flattened shell covers the animal when it is pulled down tightly against the rock. The shells are particularly handsome on the inside, and if the backs are polished, they are also colorful. The abalone meat is sold in most fine restaurants, and it is quite expensive. The shells are sold in shell shops and are used for everything from ash trays to fine jewelry.

The **periwinkles** belong to the genus *Littorina* and are probably the most common gastropod. They occur in the high tide zones of the world and are composed of many species. They are rather small snails, some of which can withstand long periods of dryness because they are able to seal their shell and consume the water within it. Some have lived over a year sealed up and out of the water, only to become active again when submerged. They can be found along most rocky areas in tide pools or high and dry in cracks in the rocks. They often fill a depression or crack in the rock with their shells.

THE CLASS BIVALVIA (PELECYPODS)

The clams, oysters, scallops, and others of this group have two shells that are located on either side of the organism. Because the shells are called valves, this group is commonly called the *bivalves*. No two species have shells exactly alike, so the shell is generally used to identify them. Most of this group are marine, with only a few fresh water representatives. Most often they have separate sexes and release sperm and eggs into the water to be carried to each other for fertilization. The author observed the male rock scallop dispersing sperm in the Gulf of California (Sea of Cortez) during a diving expedition in 1969. The scallop opened its shell slowly and then after two to three minutes closed the shell violently, causing a pulse of water to shoot out. The sperm, which could be seen as a milky fluid, was carried out by the water. Many scallops in the area were going through this sperm-broadcasting process all day long. The next day none were observed to do it. The bivalvia are not grazers and do not have the algae scraper (radula) of the gastropods. They obtain food mainly by filter feeding, and they have a siphon system to accomplish it. They have no distinguishable head or tentacles; neither is needed for their life style.

The shell is composed of three main layers. The outer layer is thin, almost skinlike material, only harder—more on the order of fingernails. This layer, called the **periostracum,** protects the second layer, which is heavy and composed of calcium carbonate, from being dissolved by the carbonic acid that occurs naturally at times in the water. The inside layer is composed of mother-of-pearl formed in very thin layers by the mantle membrane. These thin layers are responsible for the shell becoming thicker as the animal grows older. The oldest part of the shell is located at the hinge joint. This section is called the **umbo.** When we examine a shell, growth rings can be discerned as concentric lines progressing away from the umbo. The two shells are opened and closed by the adductor muscles. Some types, such as the pismo clam, have two adductor muscles, one forward of the umbo and one posterior of it. Other types, such as the pen shells and rock scallops, have one large muscle near the center of the shell. In some cases these larger adductor muscles are used for food. The next time you have a scallop dinner, you will know it is the large adductor muscle of the scallop that you are eating. Many of this group are well known as delicacies. When they are available, oysters, clams, scallops, and mussels are eaten in almost all countries.

Many bivalves are raised commercially, but the best known of these is the oyster. Extensive oyster beds are carefully tended on both the East Coast and the West Coast of the United States. The "farmers" who tend them watch mainly for any pollution that humans might put in the water, silt that could be washed in by heavy rains, and predators such as sea stars. These and other conditions that could damage the oysters are kept under close control so that the maximum yield of food can be harvested. In December 1976 some native rock oysters were found on the Pacific side of the lower tip of Baja California, that had a full pound of meat in each oyster. Because of their location isolated from humans, they had survived to grow larger than normal. The pismo clam, once found in great abundance along the sand beaches from central California south, well down the Baja California coast, was so numerous and desirable to eat that it was almost wiped out. A town in California, Pismo Beach, was even named for it.

The Bivalvia have three main life styles. Some bore into the substrate for protection; some grow on the substrate by attaching directly to it with a cement; and others attach with threadlike material known as **byssal threads.** The sea mussels are a good example of byssal thread animals. In some species the threads are soft and gold in color and were used by our ancestors to weave into cloth, the "cloths of gold" of ancient times. The large pen shell found in the Mediterranean was of particular importance for weaving cloth. Among the type that cement their shells down are oysters and scallops. All or part of one shell is made fast to the substrate, while the other shell is left free so it can open and close. This type of habitat gives good protection—from the elements, because they are very solidly attached, and from humans, since they become difficult to tell from the rock they grow on.

The burrowing types create a hole in the substrate to hide themselves and to gain protection from the environment. Some clams use their foot to

Figure 17–7   Some clams have the ability to bore into solid rock. As they grow, they enlarge the hole in the rock until only a portion of them can be seen. Here just the ends of the shell still can be seen. The majority of the shell is out of sight.

dig in the sand or mud. Others have a mechanical means of scraping a hole with their shell, or some can produce a mild acid and bore into rock by dissolving it. Because the animal starts to bore when it is very small, the borers often find themselves trapped. They extend their siphons out the small hole to the surface to feed and respire. The largest group of boring clams are the piddocks. They can be found worldwide and account for many of the holes found in rocks along the shores. They also bore into wood and are of concern to boat owners and marine builders. While the piddock is of concern, the ship worm is a disaster. It is a wormlike mollusk that bores into ships, pilings, or any wood structures. It honeycombs the wood until it collapses. The wood used now for pilings is heavily treated before use to protect it as long as possible from the "ship worm" or **teredo.**

The teredo is also responsible for a great deal of wood being ruined before it even gets to the mill. When the logs are cut, they are stored in log rafting areas in the water. A large number are rafted together (sometimes in an area the size of a football field), then a tugboat tows them to the mill.

The teredos attack the logs during the time they are stored in the water. A log stored for too long a time becomes heavily damaged.

Another economically important bivalve is the pearl oyster. This oyster creates layers of mother-of-pearl around any foreign object in its shell. The Japanese have made the creation of cultured pearls a fine art. They plant a small "seed pearl" in the shell of the pearl oyster and then care for the oyster for several years. When the seed pearl is removed, a pearl has been created by the oyster. The longer it is left, the more layers will be laid down and the larger it will be. This technique has made possible the fine-looking and relatively inexpensive pearl jewelry that is now common. The Bivalvia are an economically important class of mollusks to humans. They furnish food and jewelry, but also cause damage to structures.

THE CLASS CEPHALOPODA

Of all the animals in the sea, the cephalopods are the second most feared of all. The sharks would undoubtedly take first place, but the octopus and the squid would be second. The cephalo (head) poda (foot)-type animals are intriguing to the general public because they are uncommon, shaped differently from most other sea life we know, and have suckers to hang on to things instead of fingers. Their appearance and the stories handed down from mythology and science fiction tend to make us a bit uneasy about these animals. This reaction is unfortunate because, for the most part, they are not large, but rather timid, delicate animals. The cephalopods are the most highly developed of the mollusks. They have excellent eyesight, can swim very rapidly, show emotions, change color very rapidly by use of chromatophores, and can crawl on the bottom or swim if they choose. The group includes the octopus, squid, argonauta, cuttlefish, and the *Nautilus*.

Cephalopods are soft bodied and do not have the heavy shell that is found in most other mollusks. Their mantle wraps around the body, forming a rather loose collar in the neck region. A siphon that brings water over the gills is located under the mantle and can be used to form a jet of water that propels the animal through the water if it wishes to move rapidly. The cuttlefish and the squid have an internal shell that is not visible until dissected out. The squid shell was once used for stays in ladies' garments, while the shell of the cuttlefish is used in birdcages today. The *Nautilus* has a beautiful compartmented shell, very light in construction. The animal lives only in the last compartment. The shell is often seen in shell shops and is quite expensive. The female argonauta, or paper nautilus, has a very thin shell that is used mainly as an egg case. The male is smaller and has no shell. The octopus, with no shell at all, is atypical of the phylum in that respect. This group of cephalopods has around 400 living species, all of which are marine; but

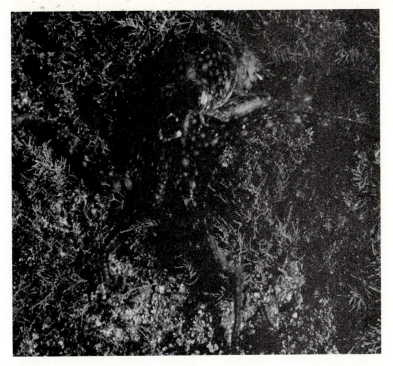

Figure 17–8   One of the more unusual mollusca is the octopus. It has no shell and can move either by crawling, as shown, or by swimming with a jet action by pushing water through its siphon tube.

our fossil records tell us that at one time there were over 10,000 species. They are an ancient group and in the past included the largest shelled invertebrates known.

The ability to change colors is one of the most interesting characteristics of the cephalopods. The color change is accomplished by muscles that pull the **chromatophores** up to the skin surface into flattened plates that show their color. The object is to match the immediate surroundings and be difficult for predators to see. They are experts at this. The octopus generally blends perfectly with its surroundings. When detected and disturbed, it will flash from one color to another, a fascinating change to watch. On a night dive off Grand Cayman Island in the Caribbean, the author and his diving companions watched for 10 minutes while an octopus, about 4 feet across, flashed almost every combination of colors we could think of. All we did was keep a light on it, which must have totally confused it. The entire book could be devoted to discussing the many interesting facts about the octopus.

The squid, like the octopus, is also very interesting. A ferocious predator, it attacks small fish as well as other organisms for food. Squid often swim in schools and were seen by Thor Heyerdahl in the Kon Tiki expedition sailing out of the water in much the same way that a "flying fish" sails above the surface. Some deep water species of squid get very large. A specimen taken off Newfoundland was over 1 meter in body diameter and over 6 meters long from head to tail. The tentacles were another 10 meters long, making the total length of over 17 meters. It was estimated to weigh over 6,000 pounds. Probably this captured specimen was not the largest in the sea; we can assume there are even bigger ones. Some investigators believe squid reach over 30 meters in length, the largest of all invertebrates. These larger squid are a prime food for the sperm whales. They also are probably responsible for stories of the mythological "**Kraken,**" a monster which is supposed to be several hundred meters across and have a thousand arms. A large school of big squid feeding on the surface with their tentacles flapping about on the surface would fit the description of the Kraken very well. The squid and octopus are also well-known for the dark brown inklike substance they shoot out when attacked. This "ink" does not function to hide the octopus, but rather numbs the sense of smell of the attacker. An eel, for instance, will not attack unless it can confirm the food source by smelling it. When an octopus shoots ink in the face of an attacking moray eel, the eel cannot locate the octopus even though the eel can see it until the sense of smell returns, which may take an hour or so.

The female octopus is a good mother. She cares for her eggs until they hatch. Because of her constant attention, almost all will hatch and very few will be lost. She keeps other animals away and is constantly cleaning them so sediment will not collect on their surface and kill them. Some species spew water from their siphon over the eggs to keep them clean; others use their suckers to lift off foreign material from the eggs. Almost all types keep any food materials or anything else that might decay away from the egg cluster. The squid, *Loligo opalescens,* off Southern California, generally schools up in January or February and deposits egg cases on the bottom. One of the famous places where this occurs is in the submarine canyon off La Jolla near **Scripps Institute of Oceanography.** Jacques Cousteau and his crew have filmed the great schools of squid for a television movie. The eggs are deposited in egg cases about 5 centimeters long and less than 1.5 centimeters in diameter. After they are exposed in the water for a few days, they swell up and increase in length and width to about double their original size. The author has had them hatch in a laboratory aquarium. Incubation takes one month, but after hatching, we were able to keep them alive for only 48 hours. Unfortunately, the time was not available to investigate why they did not survive.

The cephalopods are one of the best known, most feared, and most unusual groups in the animal kingdom. Their shape, life style, and intelligence set them apart so radically from most other creatures that they have been recognized as a special group for longer than most other biological groups.

THE CLASS SCAPHOPODA

Although the members of this class are not rare, they are not well-known to most people. They are small, live in the sand or mud buried beneath its surface, and are commonly called tooth shells. The shells shaped like long, thin fangs wash up on the beach. They are slightly curved and hollow. Both ends are open, one being larger than the other. The animal is primitive, having no heart, gills, eyes, or tentacles as most others in the phylum have. It does, however, have a shell, radula, and mantle used in shell formation, like the majority of other mollusks. Like many other shells, strings of these shells were used by the coastal Indians as money. Their relative rarity made them worth more than most other shells.

REVIEW QUESTIONS

1. Name the five common classes of mollusks, and give a distinguishing characteristic of each.
2. Why are the cowries a highly sought after collectors' shell?
3. Name the groups included in the Mollusca that do not have a shell or have it greatly reduced.
4. Describe the differences in feeding methods of each of the five classes.
5. Which class is the most highly developed in body structure and brain development?

Chapter Eighteen

THE ARTHROPODA

DEFINITION OF TERMS USED IN CHAPTER 18

Autotomy: The ability to cast off a part of the body when in danger; the part can usually be regenerated; for example, the claw of a crab.

Biramous: Having two branches.

Callinectes: Genus of commercially valuable, edible blue crab of the eastern United States.

Carapace: A hard covering protecting the back of many crustaceans as well as other animals.

Chitin: Substance that forms the shells of arthropods.

Copepods: The main animal type found in the plankton.

Crustacea: Class name of most marine arthropods.

Emerita: Genus name for the common sand crabs.

Hemocoel: Spaces in the body through which blood flows instead of in blood vessels.

Ligia and Ligyda: Indicator organisms for Zone 1 in many parts of the world; an Isopoda.

Molting: The shedding of a shell to allow a larger new one to form.

Penaeus: Genus name of East Coast commercial shrimp.

Pycnogonida: Class name of the sea spiders.

Setae: Food-gathering appendages on some copepods.

Xiphosurida: Order name of the horseshoe crab, of which *Limulus* is best known.

Zostera: Genus name for eel grass.

The Arthropoda (*arthros,* "joint"; *podos,* "foot") are the largest group in the entire animal kingdom. Their name comes from the fact that they all have jointed legs. Although this group includes the millipedes, centipedes, spiders, insects, and crustaceans, in marine biology there are only three classes of interest. The classes we will look at in more detail are the Merostomata, Pycnogonida, and Crustacea.

The general characteristics of the arthropods include a hard exoskeleton composed mainly of **chitin.** Chitin is a complex polysaccharide, a type of carbohydrate. This shell is secreted by the epidermis, and due to its restrictive nature when hardened, it must be shed at intervals to allow for growth. At each shedding a larger, oversized shell is formed, and the animal grows into the new shell. Before the molting takes place, a new soft shell membrane is formed under the shell. When the shell splits open, generally where the abdomen and thorax come together, the animal steps out of the shell, inflates the membrane with water (in the case of marine types), and when the membrane hardens in a few hours, the new shell is formed. The shell protects the animal and gives ridged areas for muscle attachment. The weight of the shell tends to restrict the size of most arthropods, and the large ones are all marine due to the buoyancy factor of being submersed.

THE CLASS MEROSTOMATA

Of the Merostomata, the order having marine types is Xiphosurida. The common name of the type is horseshoe or king crab. Of this group *Limulus* is the best known. Inhabiting the entire east coast of the United States and Mexico, it reaches a size of around 45 centimeters and can swim or walk along the bottom. It uses its flat head to dig into the sand looking for food and is active generally at night. It is an ancient form, unchanged for many thousands of years. An effort was made to transplant the species to San Francisco Bay, but *Limulus* failed to reproduce and did not survive.

THE CLASS PYCNOGONIDA

These *sea spiders* are different enough from most other arthropods that some taxonomists put them in their own phylum. They are mostly benthic marine organisms that can be found intertidal to depths of 4,000 meters. They are more prolific in polar regions, feed mostly on fleshy animals such as hydroid colonies, and share a free-swimming larval stage with the class Crustacea. Although the larva look similar, they are not homologous. The male carries the eggs cemented to his underside after he receives them from the female. The legs are normally in eight sections with a claw at the tip. The life history of this entire group is not well-known.

THE CLASS CRUSTACEA

The crustaceans, for the most part, have a **cephalothorax** (*cephalo,* "head"; thorax, "mid-part of the body"). The head and middle (thorax) are joined under a hard shell so as to be one unit and not two distinguishable ones. Mostly marine, these are the only ones we will mention. The crustaceans are a very large and significant class with a great many diversified orders. The first we will consider is the most atypical.

Figure 18–1 The sea spiders or Pycnogonida are one of the strange groups of arthropods in the sea. They are not closely related to others within the phylum.

The Subclass Cirripedia: The Barnacles

The barnacles do not at first appear to be in any way related to the crabs and shrimp, but they are. If we were to remove the shell of a barnacle, we would find a typical crustacean with its head cemented to its rock. How does it get here? Barnacles have both male and female sex organs; in scientific terminology, they are hermaphroditic. They do not fertilize their own eggs but pass sperm on to other barnacles near them through an extendible penis that can reach, in some cases, several inches. When the eggs are fertilized, they develop into a nauplius-type larva that swims into the plankton. It molts several times and changes form to become a cypris larva with a shell. It has fat droplets in the shell, which assist in flotation. It will remain planktonic for 1 to 12 weeks then, depending on the species, it sinks to the bottom and looks for a place to attach. It attaches by cementing the top of its head to a rock, boat, or some other object by use of a cement gland. After it attaches, it changes form completely (a complete metamorphosis) and secretes a calcium shell around itself.

The barnacle takes on two general shapes. One is with its shell solidly built right on the substrate. The other has a leathery stalk attached to the substrate; these are the gooseneck barnacles. Almost all barnacles can close their shell and wait out unfavorable conditions, such as drying out at low tide or keeping out fresh water during a heavy rain. They all feed in the

Figure 18-2   The bottom of an acorn barnacle removed from its rock shows the structure of its shell. The honeycomb construction gives it great strength around the central cavity where the animal lives.

same general manner, opening the shell and kicking their feet out to catch plankton that have been carried there by the current. They are filter feeders.

Because the shell-like structures they build are different, each organism can be identified quite easily. Barnacles grow almost anywhere they can get planktonic food. The best areas are inshore and on drifting material at sea. One of the more often seen is the genus **Balanus,** commonly called *acorn barnacles.* Varying a great deal from species to species, these may be 3 millimeters to 12 centimeters in diameter and as high as 15 centimeters. Because they, along with other species, will grow on the bottoms of ships, they are an expensive nuisance to clean off. Many of the barnacles have very wide ranges, one reason being that ships often carry them over long distances in a short time. Thus they are able to survive passages from good growing conditions, through poor conditions (for example, warm water), and back into good conditions—perhaps in a different ocean.

The Subclass Copepoda (Oar-foot)

Copepods, small animals just visible to the naked eye, are one of the major components of the energy cycle or food chain. They fit between primary producers, like diatoms, and small fish. Some of the large filter feeding animals, such as the blue whale, can also filter out organisms as small as the copepods. Some copepods are modified to be parasitic on fish and are called *sea lice.* These have compressed bodies, the legs shortened with hooks at the end, and the mouth modified for sucking. The genus **Calanus** is the most abundant type in the planktonic population. The planktonic species are most important because of their role in the energy cycle. There are more species of copepods than all the rest of the zooplankton, as well as more individuals. While the diatoms dominate the plant life, the copepods dominate the animal life. Their main food is diatoms, and these are generally caught by featherlike setae located near the mouth. A few types are actually predators and capture other copepods as well as other zooplankton. The population of *Calanus,* mentioned above, becomes so dense in northern Atlantic waters that it might be mistaken for red tide. It is the main food source for many fish, as well as a main source for filter feeding mammals like the baleen whales. People who make their living by fishing watch the water closely to make sure they are fishing an area rich in copepods, and small plankton samples are actually taken to confirm this fact.

The Subclass Ostracoda

Although the subclass Ostracoda has many species, they are little known and seldom seen by most observers. Few of them are pelagic, so they are

Figure 18–3 The acorn barnacle and the gooseneck barnacle are very atypical of the arthropods on first glance. A closer inspection of them shows that they do belong to the arthropods because of their body features, which are difficult to observe because of the shell they build around themselves.

less often found in plankton tows. Most live on or near the bottom, and are found in bottom samples from shallow water to over 2,500 meters. Their body is encased in a light shell resembling a clamshell in appearance. The heart can be seen beating through the semitransparent shell. A pair of eyes is located on the sides rather than on the anterior end. Equipped with rather well-developed antennae, they swim about quite easily. Whereas most other crustaceans have larval forms that go through many molts, the ostracods develop from egg to adult without metamorphoses.

The Subclass Branchiopoda

The branchiopods are a relatively insignificant group except to the aquarium keepers. They have leaflike appendages with gills along the edges. They occur mainly in fresh water and are often called *water fleas.* The main genus of interest here is *Artemia,* the **brine shrimp.** This little creature can withstand extreme changes in the salinity of its environment. They adapt from fresh water to saturated salt brine where salt is caked along the shore and are often found to be the only living inhabitant of salt ponds. They reproduce only in a certain salt concentration, higher than sea water but lower than a saturated brine solution. Normal salt evaporating ponds are perfect for them. Sold as aquarium food on a large scale, one can buy the eggs, keep them until the food is needed, and then drop them into a concentrated salt solution where they hatch. They are then used as fish food.

The Subclass Malacostraca

The subclass Malacostraca has many orders, which include the majority of the larger crustaceans. All of the orders will not be mentioned nor will any extensive review be given of any one order. For more information on any one type, the student should consult a good book on zoology.

The Order Isopoda

The isopods have a widely diversified range of habitats; they live in lakes and rivers, on land, and in the marine environment. The common characteristic is that they are flattened dorsoventrally, as if they had been stepped on. The land isopod, called the *sow bug* or *pill bug,* is a good example. This group contains several species, commonly called *gribbles,* that bore into wood. They destroy pilings, wharfs, and boats. Because of these and other boring creatures, the wood used for marine use must be protected. These organisms are one of the reasons why boats must be hauled out of the water each year and painted with a poisonous antifouling paint, which keeps the organisms out of the wood for a year or so. The genera *Ligia* and *Ligyda* are

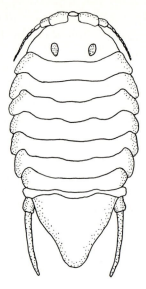

Figure 18–4 The isopod *Ligia* is one of the most common isopods. This is a general drawing showing the flatend body, which is characteristic.

common types in most of the world. They have adapted to a life out of water along the edge of the sea and are so well adapted to life in the air that they will drown if held underwater for an extended time. One group of isopods is parasitic on other crustaceans, but not on any other organism. They are a widely diversified and interesting group.

The Order Amphipoda

The amphipods are similar in many ways to the isopods in that they both have a segmented thorax and not a cephalothorax. The thorax is divided into seven freely movable segments; this allows them more movement. Several of the common amphipods are called **sand hoppers** or **beach fleas.** Where the isopods were flattened from top to bottom, the amphipods are

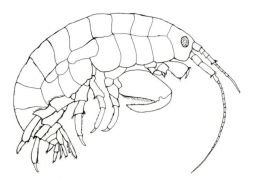

Figure 18–5 There are many species of Amphipods along the beach and in the intertidal zones.

flattened laterally from side to side. The sand hoppers have made the move to land much as *Ligia,* the isopod, has. In one genus found in Europe, *Orchestia,* there is a species whose normal habitat is the upper beach. Most of the hoppers live in the sand and around debris on the beach just above the high tide level. During the day, they burrow in the sand, but at night they come out to feed at low tide on the material washed up and left on the beach during high tide. Consequently, they must relate to two cycles, day-night and high-low tides. They become very active feeders when a low tide occurs at night. They do not like to be disturbed and are not often seen by daylight because of their burrowing habit. Some species weave a weblike tube for protection, while another sews the edge of a kelp plant over into a tube to hide in. Members of the hoppers grow as large as several centimeters and are important beach cleaning organisms.

Not all amphipods live on beaches. Many live in tubes they form with mucus and mud particles on the bottom. Some of these forms crawl or swim across the bottom and draw their tube along much the same as a snail carries its house. Others stay on the bottom during the day and then swim up to join with other plankton during the night. Some amphipods are parts of the holoplankton and are significant in the zooplankton population. One of the more interesting planktonic habitats used by some amphipods is the jellyfish. The amphipods live either as a parasite or symbiotically with many jellyfish and some salps (a tunicate).

The most unusual group of amphipods is the caprellids. In many ways they remind the observer of a praying mantis. They are long (up to 2.5 centimeters) and very slender. They grasp the weed on which they live and reach out to catch food. Commonly found with hydroids, they are difficult to see because they blend so well with their environment. Much like a worm humping along, they move by reaching out and grasping the substrate with the anterior legs, releasing the posterior legs and humping up to regrasp with the hind legs again. The caprellids include one genus that is an exception to the side-to-side flattening of most amphipods, the genus *Cyamus,* called the *whale louse.* It is parasitic on whales and is flattened dorsoventrally like an isopod.

The Order Stomatopoda (Mantis-Shrimps)

Stomatopods are found in burrows in the sea floor. They burrow down to over 1 meter and maintain a hole of 5 to 10 centimeters in diameter. They have a set of jackknife claws that they can flip out with such speed and accuracy as to cut other small animals in half. Some types grow to 30 centimeters in length and are quite dangerous to handle. They are fine eating, but rarely caught in large enough numbers to be a significant food source.

The Order Euphausiacea (Krill)

This group of small (up to 6 centimeters) shrimplike crustaceans can be identified by the following characteristics: compound eyes located generally on
stalks; light organs, which are often found near the base of the eyes as well
as on the body; and gills attached to their legs, not covered by the carapace
that covers the first eight segments of the animal. Their **biramous** (two-
branched) legs are another easily recognizable characteristic. Some are
nearly transparent; others are bright red. Their main role in the food chain
is for large filter feeders such as the whalebone whales. Gathering in gigantic
populations, krill are easy prey for these giant mammals. The Antarctic environment is particularly rich in euphausiids, a main reason there was a large
whale fishery in Antarctic waters. Most of the whales have been killed off
by overfishing and are now gone. It was not uncommon while traveling in
Antarctic waters to pass through a red section of water with whales and
penguins feeding voraciously on the surface. In 1958 during the International Geophysical Year, the author observed such a feeding pattern in the
Ross Sea during the United States Antarctic expedition. The species found
there, (*Euphausia crystallorophias*) is red and around 6 centimeters long.
They were easily observed and a joy to watch as they darted just under the
surface.

The Order Decapoda (Ten Feet)

The decapods include the best-known crustaceans. People who have no
knowledge of the sea still recognize lobster, shrimp, and crabs; their edibility

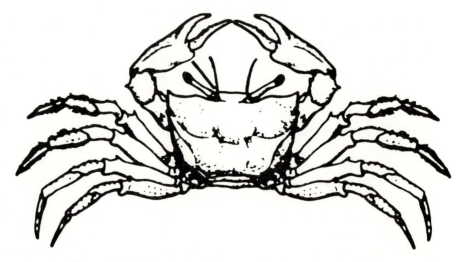

Figure 18–6 A main characteristic of arthropods is jointed legs.

has made their reputation worldwide. As a group, they are economically as well as biologically important.

Decapods are characterized by five pairs of walking legs. The order is divided into the organisms with an elongated abdomen, such as the shrimp, prawns, lobsters, and the ones with a short abdomen, such as the various crabs.

Shrimps.   Shrimps have been taken as food for over a thousand years. As far back as 1911, shrimping was outlawed in San Francisco Bay due to overfishing, but was later reinstated under controls. The main shrimp fishery is located along the Atlantic coast and in the Gulf of Mexico; however, there is a fair-sized fishery off the western coast of Mexico and Baja California and a small one in the Pacific Northwest. All shrimp are not the large edible ones. There are many small, delicate types that inhabit tide pools and other areas. One of these tide pool shrimps, *Spirontocaris,* is transparent, and the observer can watch most of the internal organs work. As in most animal groups, the animal adapts to environmental conditions by the modifications it shows. The shrimp that lives in the eel grass (*Zostera*) is green; as it clings to a blade of eel grass, it is almost impossible to see. In the tide pools of Southern California several small shrimps will respond to being fed by coming out and feeding right under the nose of the observer. Scuba divers often play with these species in the shallow, rocky areas. The most common of these is probably *Hippolysmata,* a red shrimp with broken lines or stripes running from the head to the tail. These small, delicate types have no commercial value and are only seen by the astute and interested observer. The genus **Penaeus,** on the other hand, is the main food shrimp caught on the East Coast of the United States. It is well-known. Extensive information is available on these commercial species through fish and wildlife studies of the Gulf states and also from the federal government. Some of the small shrimps also have large claws. One of these (*Betaeus*) has a set of large claws and is often mistaken for an East Coast baby lobster. It is found under the shells of abalone by divers. Another type, called the "pistol shrimp," has one large claw and one small one. The large one has a small fingerlike snapper on the end that can be snapped with such force that the vibration knocks out small animals near it. This is how the shrimp catches its food. Divers can often hear the cracking of the pistol shrimp. In still water at night, the noise they produce vibrates through the hulls of boats and can be quite annoying to a tired sailor trying to get some sleep.

Lobsters.   There are two lobsters of importance: the American lobster, **Homarus,** of the East Coast, with its gigantic claws, and the spiny lobster of the West coast, **Panulirus,** which totally lacks claws. The large claws of the American lobster make up to one-half of its body weight. The larger the

Figure 18–7   The lobster is one of the most prized of all the arthropods for food. Lobsters support a substantial commercial fishery worldwide.

Figure 18–8   A small arthropod has residence in an old mussel shell.

lobster is, the larger the percentage of weight found in its claws. A 35-pound lobster recorded by Herrick was over two-thirds claws by weight. Practically the entire lobster is edible except the shell and a few internal organs like the gills and green glands (excretory glands). The female lays her eggs generally in July or August and carries them tucked under her tail for around ten months. A specimen of about 34 centimeters (not counting the claws) will lay approximately 30,000 eggs. After hatching, they swim for six to eight weeks then settle to the bottom as their metamorphosis continues and they take on the general appearance of an adult.

The spiny lobster of the West Coast has no claws, and its main defense is the spines on its back and along its antennae. A fish that gets too close has the spiny antennae rubbed across its eyes and quickly backs off. The female carries her eggs only a little over two months, and the larvae settle to the bottom after several weeks to mature. The top weight is about the same as the American lobster, around 35 pounds, except that all of the weight is in the body. The regenerative powers of the lobster are truly amazing. All ten legs and both antennae can be removed and the animal will survive. It will molt in about six months and have all new parts. The new parts will be complete, but miniature. It takes several molts to regain the full size of the limbs. The spiny lobsters (several species) are found in most tropical areas from Florida to Central America and in some areas of the South Pacific.

The True Crabs (Suborder Brachyura)

In the true crabs, the abdomen has been modified to a smaller flap generally tucked up under the body. The ten legs, characteristic of the decapods, are plainly visible, with the first set generally having claws. Male and female can often be told apart because the abdominal flap is **narrow** on the male and **wider** on the female. In the spider crabs the male will generally have large claws and the female small ones. The crabs are a large group and rather diversified. Most are scavengers, but they will capture live food if they can although they are not generally fast enough to catch much. They walk sideways, with four feet pulling while the four on the other side push. Underwater they move easily because of the buoyancy of the water; on land they move with much more difficulty because they need to support their rather heavy body. Because their eyes are set up on stalks like those of many other crustaceans, they have the great advantage of being able to see 360° around. Nothing can sneak up on them. The legs are modified in some species: the fiddler crab has one claw much larger than the other, and the swimming crab has the last two legs modified into paddles for swimming. The gills are attached at the base of the walking legs and move as the crab moves its legs, circulating water over them. The carapace extends to the sides to cover the gills and protect them; consequently the crabs often have bodies wider than they are long. Most of the crabs are considered to be fine eating, but only a few types are taken commercially.

The Spider Crabs.   Spider crabs are characterized by having a sharp point on the front of their shell and long, slender legs. They are the largest of all the crustaceans. One species of spider crab that lives in the Pacific Ocean grows to over four meters across. Others are small and delicate, like the decorator crab. This small crab gets its name from the habit of placing whatever it can find on its back for camouflage, most commonly bryozoans, sponges, and seaweed. Some of the larger crabs show the decorating or masking instinct only until they are large enough to protect themselves. Once they are large, they can fight quite well with their quick claws and no longer need the protective decoration. Many species of spider crabs live in the kelp, and these are commonly called *kelp crabs.* They have the long rostrum or nose and are easily identified as spider crabs.

The Swimming Crabs.   The most important of all the swimming crabs is **Callinectes.** This is known as the edible crab or blue crab of the East Coast. It is known to be a highly prized delicacy in restaurants of the eastern seaboard. Its close relative, the lady crab, is also used as food. On the southern part of the West Coast and Mexico, there is another swimming crab, *Portunus. Portunus* is not a food crab, although it is often used as fish bait. The members of this group have the last pair of legs modified into paddles for swimming and are very good at it.

Cancer Crabs.   While the blue crab is the important commercial crab of the East Coast, the cancer crabs are the important ones on the West Coast. They are also found on the East Coast, north of Cape Cod. The average cancer crab taken for food is from 15 to 21 centimeters across the carapace, although they have been found up to 40 centimeters. They are commonly taken from rock jetties with open hoop nets or folding traps by sport anglers.

The Pea Crabs.   The pea crabs are small crabs that normally live in a commensal relationship with something else. They are often found in clams, mussels, or other molluscs. They live in or under the shell, deriving protection from it and extracting food from that gathered by the host. Only one individual seems to inhabit a given host, and thus they do not become detrimental. They also are common in the tubes or burrows of various worms and shrimps and on the shells of sand dollars. They often have modified shapes to fit their various accommodations.

The Shore Crabs (Grapsoid Crabs).   These shore crabs have a squared carapace, with their eyes widely separated on their two front stalks. They run rapidly across the sand or over the rocks and are highly visible to the public. The "**ghost crab**" (*Ocypoda*) is a beach burrower on the East Coast, whereas the **striped shore crab** (*Pachygrapsus*) is a rock crawler on the West Coast. These two genera are the most common of the many shore crabs.

The Fiddler Crabs.   If all things in the animal world were as easy to identify as a fiddler crab, how wonderful it would be. The male of these small mud-flat organisms has one gigantic claw and one very tiny claw. When they move the claws back and forth, as they do when they are threatened, it looks as if they are playing a violin. They eat organic material out of the mud, creating small mud balls as they finish. These are taken out of their burrow. These interesting little creatures were a common organism in most bays on both coasts years ago. Now that there are pollutants in the bays and many of their mud flats have been made into marinas or housing tracts, they are not doing very well.

The crustaceans are an extremely widespread and visible group. We have seen how economically important they are and how they have adapted to fill habitats in all environments. Because in most cases their eggs hatch and the larvae enter the plankton for a period of a few days to a few months, this process is very significant to the planktonic feeders. A crab may produce 30,000 larvae, of which three may survive. The rest have been used as food for other animals. Some will be eaten alive; others will die and become organic material in the mud for the worms, but none are wasted. The energy cycle continues to utilize all of the energy available to it in one way or another. Most of the crustaceans undergo metamorphosis and before becoming adults go through one or more developmental stages known as nauplius, cypris, protozoea, zoea, mysis, etc.

Other Crabs

The Hermit Crabs.   Hermit crabs are decapods that have rather soft unprotected abdominal areas and for protection find empty snail shells to live in. Some large land hermit crabs carry about a 2- or 3-pound shell, but most of them are small and inhabit shells of 1 centimeter to 7 centimeters in diameter. Their bodies are curled up and offset to the right. When they pull back into their borrowed shell, they are quite well protected. As they grow, they just find a larger shell to move into. They can be seen running around quite actively in many tide pool areas and they also have been collected from deep water. The hermits, like the lobsters, are scavengers: They will eat almost anything they can catch or find on the bottom. They are mainly meat eaters, although on occasion they have been observed eating algae. The hermit crabs seem to be ancestral to some other forms of crabs, such as the stone crab and the coconut crab, which has left the water entirely and lives on land. Its food is basically coconuts; it will even climb trees to get them.

Figure 18–9   The hermit crab finds empty snail shells to live in. When it outgrows the shell, it finds a larger one and moves in.

The Sand Crabs.   The sand crabs are not considered true crabs. They are transitional between the types with large abdomens, such as lobster, and the true crabs with small abdomens. The sand crabs, along with hermit crabs, porcelain crabs, and several others, are placed in the transitional grouping of Anomura, which is sometimes called a *tribe.* Sand crabs are well-known to the beach angler; they are easy to catch and are a choice bait for several types of surf and rock fish. The person fishing will spot them when they dig in as a wave recedes off the beach. They often leave their eyes, which are on stalks, and their long antennules just out of the sand. As the water flows back off the beach, it makes a V-shape ripple as it passes over them. It is easy to reach a hand down and scoop them out of the sand. Their shape is modified so they can swim or dig into the sand. A hard, elongated carapace covers the entire oval-shaped body and protects them from the abrasive action of the sand in the surf line where they live. They are filter feeders, leaving their feathery antennae out in the water when they burrow in. Their burrowing is done tail first. Quickly they dig into the sand in a second or two so as not to be washed out to sea.

Because this environment of sand and surf is one of the harshest environments in the sea, the sand crabs have made many adaptations in their structure to survive. They have a smooth, hard shell and use their antennules for respiration, their legs as digging tools, and their antenna to trap small organisms for food. Where the hermit crabs are the entertainers of the tide pools, the sand crabs are the entertainers of the surf zone along sandy beaches. An observer can watch them move up the beach with an incoming tide, and down the beach with an outgoing tide. They stay in the zone of maximum exposure to receding water so they can feed. Of their many species, most are in the genus *Emerita*. The most numerous species is *Emerita talpoida* along the East Coast and *Emerita analoga* along the West Coast. When the female is about to lay eggs, several males will attach to her by use of suction cups located on their fourth leg and deposit sperm on her underside in the form of a mucous ribbon. The female lays her eggs, uses the sperm to fertilize them, and then carries them for up to five months. The larvae join the plankton for a while as they molt, and then settle to the bottom. If the water is warm and conditions good, they will remain active all year. If the weather on land is very cold or ice forms on the beach, they stay in deeper water.

Another member of this group, sometimes called the spiny sand crab or mole crab because of its shape and burrowing habit, is *Blepharipoda occidentalis*. This is a larger animal, reaching 5 to 8 centimeters in length and living in deeper water behind the surf zone rather than in it. The shell has sharp spines on it. They are not commonly found although they may be more abundant than was previously thought. Scuba divers often find them along the Southern California coast while digging for pismo clams. Often something is thought rare until a new or different method of sampling is used, and we find the organism was not rare but just undetected.

The Porcelain Crabs.   The porcelain crabs are mainly inhabitants of rocky areas. They often live in pairs of male and female and have a flattened body so they can fit into small crevices. They have large claws that they use for protection and picking food. These claws are quickly dropped off the body if they are seized. This ability to quickly shed a part of the body is called **autotomy.** It is a means of survival, as it allows the animal to escape if caught by a claw or leg. Members of this group are found in rocks, kelp, and mussel beds or anywhere they are well protected. They are also members of the tribe anomura, so they are not considered true crabs.

REVIEW QUESTIONS

1. Why are barnacles included in the same class as crabs?
2. Why are copepods so essential to the marine environment?
3. In what way are the isopods "gribbles" economically important?
4. The order Euphausiacea is important to what other group of animals as food? Where is it most concentrated?
5. Name six different types of true crabs, and give some point of interest about each one.

Chapter Nineteen

ECHINODERMATA

DEFINITION OF TERMS USED IN CHAPTER 19

Aristotle's lantern: Common term used to describe the way the teeth are assembled in most sea urchins.

Asteroidea: Class containing the sea stars.

Crinoidea: Class containing the sea lilies and feather stars.

Echinoidea: Class containing the sea urchins and sand dollars.

Holothuroidea: Class containing the sea cucumbers.

Ophiuroidea: Class containing the brittle or serpent stars.

Ossicles: Part of the structure of the sea star skeleton.

Pedicellariae: Small pinchers found on the back of many echinoderms used to keep the skin clean as well as sometimes to catch food.

Respiratory tree: The major respiratory organ of the sea cucumbers.

Test: A word used to designate the shell of the sea urchin.

Trepang: Dried sea cucumber used as food.

Tube feet: A part of the water vascular system found only in echinoderms; used mainly for movement and catching food.

The echinoderms have always been one of the sea's unique groups. When humans first classified the echinoderms, they were placed with the Cnidaria because of their radial symmetry. They were given a name, the Radiata, based on this one characteristic. As our biological knowledge increased through research, the differences between the two groups became obvious. In fact, about the only thing these two groups share is radial symmetry, and in echinoderms this is an adult trait; larval symmetry is bilateral. Their radial symmetry separates both groups from the great majority of the rest of the animal world, but they differ from each other because the echinoderms have a complete digestive system with a mouth, intestine, and anus. They also have a **water vascular system,** found only in echinoderms. The general characteristics of the phylum are a water vascular system, solely marine distribution, and a skeleton composed of calcareous plates that lie under their skin. Their radial symmetry, when present, gives them the advantage of encountering their environment equally as well from any direction. Almost all forms are benthic, so their substrate protects them from attack from below and their spiny skins, spines, or leather-like hide, depending on their class within the phylum, protect them from attack on their upper side.

The water vascular system is unique to echinoderms and worthy of more explanation, because it is very apparent when any sea star is handled and examined. On the underside of the sea star, are many tube feet that move in and out and have suckers on the ends to hold on to what they touch. They vary in number from a few to hundreds, depending on the size and type of animal, but in all cases, the tube feet are used for locomotion and feeding. The general construction of the system can be seen when the animal is dissected. It is a good lab exercise. The water vascular system in the sea star consists of a small opening in the dorsal side with a sievelike plate covering it. This is the **madreporite,** the initial opening for water to enter the system. It is plainly visible on most sea stars as a white spot on the top of the central section. The water passes through the **stone canal,** a short vessel between the madreporite and the **ring canal,** a circular vessel around the center part of the sea star. Each arm of the star has a **radial canal** that starts at the tip, runs the length of the arm, and connects with the ring canal in the central portion of the animal. Along this radial canal the **tube feet** are located. The tube feet can be extended or contracted by muscle action in a balloonlike structure on the internal end. The structure, called an **ampulla,** can enlarge and suck the water out of the tube foot to contract it or contract and push water into the tube foot to elongate it. Each ampulla controls the action of its own tube foot. A sea star has many tube feet, an advantage when it feeds on bivalves. Crawling over a bivalve (clam, oyster, etc.) and locating it in the center just under the mouth, the sea star brings its arms down over the shell, attaches the tube feet with suction, and starts to pull the shell apart. The shell has one or two large adductor muscles that hold

Figure 19–1   Tube feet are a characteristic of echinoderms, shown here on the underside of the sea star.

Figure 19–2   These three sea stars show the variety that can be found even in a single tidal pool. These were taken from a tide pool in Baja California.

its shell closed. The many tube feet pull steadily against the much stronger adductor. When one tube foot tires, it is relaxed and rested while another one takes its place. By rotating the feet, the sea star keeps a steady pressure on the adductor muscle of the mollusk until it tires enough that it can no longer hold the shell closed. Depending on the size and type of bivalve and the size and type of sea star, this tug-of-war to open the shell will last from an hour to a week. The sea star nearly always wins.

The echinoderms are generally divided into four classes. Stelleroidea is the class and Asteroidea the subclass that we normally call sea stars. The subclass Ophiuroidea are sea stars with a dishlike body and very slender arms, generally called brittle or serpent stars. The class Crinoidea, unlike the other two star types, have a mouth turned up and are called, among many other common names, sea lilies and feather stars. Two classes are not sea stars of any sort and, to the casual observer, very little resemble the members of the two classes already mentioned. The sea urchins and sand dollars belong to the class called Echinoidea and have a shell (or **test,** as it is called in sea urchins). The other class has a tough skin and no shell. The sea cucumbers belong to this class, called Holothuroidea. Representatives of the phylum and, indeed, of most of the classes are found around the world in most seas and at most depths.

A new class of echinoderm was found in 1985 by a researcher at The Australian Museum. It was found feeding on bacteria in waterlogged wood at a depth of 1,000 meters. Although this particular echinoderm is of no major economic importance, its discovery proves that major groups of life are yet to be discovered on earth.

THE SUBCLASS ASTEROIDEA

The sea stars or starfish are recognized by almost everyone. From their central part called the disc, five or more arms radiate in all directions, similar to a child's drawing of the sun. The mouth is on the underside in the center of the disc, and the anus is on the top side. Near the anus is the sieve opening for the water vascular system, the madreporite. On the underside or oral side, a deep groove runs from the mouth area out each arm. This groove, called the **ambulacral groove,** is where the large tube feet are located. The tube feet extend from the mouth area along the groove to the tip of the arm in two to four rows. At the end of each arm is an eyespot, or light sensitive organ. The general body area, especially the top side, has projections in some species long enough to call spines; in others, just long enough to make the skin rough. The projections are parts of the bony plates that make up the **endoskeleton.** These calcareous plates or ossicles are of different sizes and shapes, and each species has its own definite pattern. The ossicles are joined together by muscle fibers and other connective tissues, so the animal is flexible. The outer skin has many cilia, gills, and small pinchers called **pedicellariae,** which are very small and either difficult or impossible to see without a magnifying glass or microscope.

The number of arms varies from species to species. The smallest number of arms is 4, and the largest number is 40. The relationship of the arm to the central disk also varies greatly. Some species have very short, wide arms, while others have long, slender ones. In all cases, however, the arm blends into the central disk.

The small pinchers, or pedicellariae, are not found on all species but are common on most. They act to keep the exposed parts of the sea star clear. If a small larval type lands on the top of a sea star, it is quickly crushed so it can't grow there. Although the pedicellariae are small, often a great number of them are present. They have been observed to catch and hold small crabs up to 2.5 centimeters long by grasping the hairs on the legs of the crab. Although their main function is to keep the back of the animal clean so the gills can function, they also carry a very small amount of food to the sea star by trapping it and then slowly passing it on to the ambulacral groove by way of the cilia on the dorsal surface.

The water vascular system is one of the more interesting and important features of this class, as mentioned above. To summarize, it consists of the madreporite on the dorsal surface that takes water into the stone canal which leads to the ring canal. From the ring canal the water goes to the radial canals, which lead off into each arm above the ambulacral groove. The water then passes from the radial canal into lateral canals that lead to the tube feet. There may be hundreds of lateral canals and tube feet.

The digestive system is simple but in some cases rather extraordinary. A number of the sea stars can extrude a portion of their stomach. Consequently, when they get a clam or mussel partially open, they can extrude their stomach out of their mouth and into the clam to digest it. Others lay their stomach over a coral formation and digest the coral animal right out of its home. The best known of the coral eaters, the Crown of Thorns sea star, has become a real problem on the coral reefs of the western Pacific region. A large-scale program has been underway for many years now to kill the Crown of Thorns sea star so they will not destroy the reefs. Many methods have been tried, most of them unsuccessfully. One technique uses a hypodermic that injects formaldehyde into the star, killing it. The difficult part of every method is that a diver in the water must make contact with each individual sea star. Although one gastropod mollusk is a natural enemy of the Crown of Thorns, biologists were of the opinion that the natural control of the sea star population by the gastropod would be too slow and the coral reefs would all be destroyed. Whether or not the intervention of humans in this natural cycle is good or bad for the ecology of the area has been extensively debated on both sides. A hundred years from now our descendants will be able to tell if what we have done was wise or not. We will never know with absolute certainty.

Not all sea stars extrude their stomachs. The sand star of the Pacific coast of the United States crawls across the sand and takes in whole gastropods. It will ingest any small snail-type creature it can catch. It then spits out the empty shell through its mouth. The empty shells are often picked up by hermit crabs and recycled by nature.

Reproduction in sea stars is varied, but except for a few cases the sexes are separate. Generally, the egg and sperm are released into the water and unite there. A series of changes then occur, and a ciliated larva is formed. Generally, the larval stages remain near the bottom and in most species do not rise and become part of the surface plankton. It is interesting and of major biological importance that one of the stages through which the sea star larva passes is called a **bipinnaria larva;** at that stage it is bilaterally symmetrical, as are most of the higher forms of life. It continues to develop into its radial symmetry only as an adult. Some types hold their eggs around the mouth area and protect them, but most species release the eggs into the water to free float until they settle on the bottom to mature.

The majority of sea stars can regenerate lost arms with no difficulty. If the arm is cut off, the sea star will grow a new one. Some species have such great regenerative powers that the arm broken off will also regenerate into a whole new sea star. Many early attempts to eradicate sea stars from oyster, mussel, or scallop beds were near disasters because the sea stars were just cut in half and thrown back into the water. This, of course, doubled the population in less than six months and increased the problem. Humans have been guilty of many such follies throughout history when they have interfered with nature.

THE SUBCLASS OPHIUROIDEA

The frequently encountered brittle or serpent stars and the infrequently encountered basket stars comprise the subclass Ophiuroidea, thought to be the largest group of echinoderms. Its members are delicate and spend their lives either well protected or in quiet waters. Near shore in tide pools, one generally must turn over rocks to find them. On coral reefs they can be seen in nearly all the cracks and crevices, and in the deeper ocean they are found on the bottom or buried in the soft substrate.

The Ophiuroidea are very common, but one must know how and where to look for them. They are easy to recognize, for they have a body shaped somewhat like a coin, roundish and flattened, with the arms radiating from the mouth on the underside. Unlike the sea stars, the body and

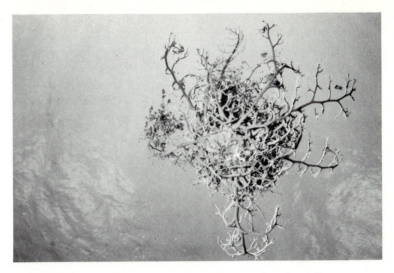

Figure 19-3 This basket star shows how its legs branch into many branches, somewhat resembling a bush.

arms do not mold together and are joined in such a way they can break off and regenerate easily in most species. The arms are slender and rather fast moving, allowing the animal to move across the bottom and, in some species, even swim through the water. Because of the high degree of flexibility in their arms and their ability to move, the tube feet are not used generally in locomotion and are usually reduced to smaller sensory and respiratory organs. Most tube feet do not have suckers, although some do and aid in passing food to the mouth. They generally feed either by lifting the arms up into the water to catch plankton or by finding dead material on the bottom. Many forms have a mucus secretion on the arms, so drifting organisms will stick on them and can then be passed to the mouth. The mouth has five plates but does not have the "Aristotle's lantern" mechanism described below that the sea urchins have, and there is no anus. Food not digested is spit back out the mouth. Feeding is done mostly at night in shallow water, and scuba divers, when diving at night with the aid of underwater lights, often see large numbers of them with their arms extended in a feeding attitude.

The subclass Ophiuroidea is divided into two groups, based on the types of arms they have. If the arms are simple and unbranched, they belong to the group known as **brittle stars.** If the arms are many branched, they are called **basket stars.** With many branches on each arm basket stars often look as if they are sitting on the bottom. This bushlike set of arms creates, with the help of some mucus, a very effective filter for extracting plankton from the water. The arms curl into the mouth and then back out again to catch more food. Watching a large basket star feed is something never forgotten. They move their arms with such grace that they look like some creature out of a science fiction story trying to wave at you.

Like most other echinoderms, they reproduce with separate sexes, releasing eggs and sperm into the water for fertilization. A few species carry their eggs in a brood pouch, but most of the larval forms are planktonic for a while before they mature.

THE CLASS CRINOIDEA

The Crinoidea are called *sea lilies* and *feather stars.* Not well-known to most people because they are mainly from deeper water, with the advent of scuba diving, they are commonly seen in many coral reef areas. Although rarely over 40 centimeters, they are often brightly colored and stand out in the environment. They consist of a central disk with five arms coming from it. Each arm divides into two or more branches, and each branch or segment has small cross-branches called *pinnules.* These branchings give the appearance of a feathery structure. The center disk is similar in shape to a cup, with the mouth in the bottom. The arms bring in small planktonic food by passing it along the ambulacral grooves with moving cilia. A stalk grows

Figure 19–4 The crinoids are filter feeders. Here we have a portion of one arm that shows the large surface area to catch plankton which is made possible by the many side projections.

from the disk and often attaches the animal to the bottom. Consequently, the mouth stays turned up, while the featherlike arms create a netlike apparatus to catch and transport food to the mouth. Some forms have no stalk, or lose it as adults, and can move their arms and swim to a new location. They are delicate forms but have a high regenerative ability to repair themselves if they are injured.

THE CLASS ECHINOIDEA

Included in this subdivision of Echinodermata are the sea urchins, sand dollars, and heart urchins. Their general characteristics are rounded bodies without arms; spines covering the body (long on sea urchins, short in the sand dollars); and a body protected by a structure called a *test*. The test is composed of sections of plates fused together to create a solid shell-like case to live in. There are generally ten double rows of these plates, with five pairs perforated for slender tube feet to come through the test. These perforations are plainly visible on the test of a sea urchin. If the sea urchin is broken open, the lines of the plates are clearly delineated on the inside of the test. The regular sea urchins and sand dollars clearly have radial symmetry as adults, but the irregular heart urchin is a borderline case between radial and bilateral symmetry. It has a slightly elongated test, with the mouth at one end and the anus at the other, and moves in the direction of the mouth. The sea urchins can and do move in any direction without a "front" part to the shell. The author has observed thousands of sea urchins moving across the bottom in a shallow sand bay on Cozumel Island off the east coast of Mexico. When food was dropped into the path of the migration, the urchins would change their direction of movement but would not rotate their bodies, thus indicating they have no special anterior or front end. This ability to face the environment equally from all sides is a major advantage of radial symmetry.

The mouth of the sea urchins and sand dollars is located on the bottom and in the center of the animal. The mouth parts or teeth are fitted together and held by a series of calcareous parts to form an easily recognizable structure. Aristotle in ancient Greece likened this structure to a lantern, and to this day it is still referred to as **Aristotle's lantern.** Because the teeth do not digest easily, they are often found in the stomachs of animals and are used to help identify the type of food that animal eats, as well as some of the predators of the urchins. The rest of the digestive system consists of a relatively long intestine with a wider "stomach" portion and the anus located on the top side. An exception is the heart urchin, where the anus is located near the margin between the dorsal and ventral sides opposite the mouth. The most unusual thing about the digestive system is a tube called the **si-**

Figure 19-5 The mouth of the sea urchin is located on the underside and has five small teeth. The mouth parts are commonly referred to as "Aristotle's lantern."

Figure 19-6 The sea urchins, as a group, are recognized by the roundish body with many spines protruding from it.

phon that starts near the mouth. It branches out of the intestine and by-passes the stomach to enter the intestine again past the stomach. It seems to function to keep water flowing through the intestine without too much disturbance of the material digesting in the stomach region.

The water vascular system is similar to that in the sea stars. The tube feet are used for movement along with the spines. The spines are attached by muscle fibers and can be moved at will by the urchins. The urchins depend on a combination of moving spines and tube feet for locomotion.

The sexes are separate, and the eggs of the larger sea urchins are quite good eating. These eggs are in wedges that resemble a section of an orange. In many parts of the world, they are a part of the commercial fishing industry. They are most heavily consumed in Italy and Japan.

The Echinoidea have pedicellariae with three jaws as well as several other types, all occurring at one time on one specimen. These small pinchers keep the surface clean and protect the urchin from enemies. On some species, the pedicellariae are poisonous. With the spines and the pedicellariae, the urchins are quite well protected. The sand dollars protect themselves by digging under the sand, while the heart urchins dig under the mud. Many sand dollars have holes, sometimes very large, that go clear through the test. These holes tend to strengthen the test, enabling it to withstand surf to a higher degree. The general shapes of this group are all similar, and the sand dollar and heart urchin are more or less modified sea urchins. Both the heart urchin, with a slightly flattened test, and the sand dollar, with its test completely flattened out, look like a sea urchin that some big fellow stepped on. Each shape is slightly adapted to its environment. The sand dollar's flat shape

Figure 19–7   The sea cucumber, with its tough skin and changeable body shape, is common worldwide in a variety of forms.

allows it to dig in the sand and orient itself in line with the surge of the surf so as to have as little drag from the water as possible. The heart urchin, with its elongated test and mouth at one end and anus at the other, is able to crawl through the mud and scoop organic material from the mud as it goes. The mud passes through the body and out the anus. Even the spines are all laid back pointing toward the anus so as to create little drag as the animal moves through the mud. Although all heart urchins are not found in mud environments, they are best adapted to mud type substrata. The larger-spined sea urchins with their rounded bodies stay back in crevices for protection in a surf zone, but wander across the bottom freely where the surf is not a problem. Their long, sharp spines radiating in all directions afford them good protection from almost everything.

THE CLASS HOLOTHUROIDEA

The sea cucumbers are, for the most part, less known by the average person than the sea stars and sea urchins because they are not commonly intertidal and do not have a shell that can be collected. Although some are used for food, they are not commonly eaten. Their appearance does not invite handling by most people, so even the divers who come in contact with them on a regular basis do not generally disturb them.

Like the rest of its phylum, sea cucumbers have identifiable regions along their elongated body, which looks much like a large cucumber. The five areas run from the mouth on one end to the anus on the other. Generally, the tube feet on the three ventral sections are used for locomotion and have suction cups like those of the sea star. There also can be tube feet along the two dorsal sections, but they are generally used for respiration or for touch. Unlike Echinoidea and Asteroidea, sea cucumbers do not have pedicellariae and spines, but they do have tube feet-type tentacles around the mouth, similar to the other forms. The body of the sea cucumber is leathery and can extend rather long or contract into a shorter animal, depending on the use of five muscles that run from mouth to anus. When disturbed, the sea cucumber draws up as small as it can and becomes very hard. When it is relaxed, it has a long, soft body. Small calcium ossicles embedded in the skin make the animal harder when it is drawn up tightly. Because the ossicles are varied shapes in different parts of the same animal, they are difficult to use as identifying traits except for an expert.

The respiration process in this class is interesting. The majority of Holothuroidea breathe through the **respiratory tree,** a branched organ of many tubes. The rectal area and the cloaca are expanded and contracted to suck water into the anus and force it up the respiratory tree tubes. Oxygen is extracted, and some water is allowed to pass through the tubes into the body and help give the body shape and form by use of water pressure.

These creatures have an ability that can be quite startling to the unsuspecting collector. When things are going badly, such as when they are being handled roughly or when they are left in water that starts to become stale, most of them have the ability to eject a large portion of their innards—in some species, out of the anus, or in some others, out the mouth. Still other species contract violently and break their bodies up into several pieces, or rupture the skin and cast out some of their organs. All of them have the ability to regenerate the lost organ. Off the island of Cozumel, the author grabbed and yanked one specimen over three feet in length that was attached at one end in the coral reef. It was during a night dive, and the animal was stretched out feeding. When it was yanked on, the entrails were cast out. They were in the form of a stringy mass that was very adhesive. It stuck to both the diver's hands, and when his buddy came over to help him pull if off, it stuck to him also. The sticky filaments could be removed only after the divers left the water. This type of protective device would certainly discourage an attacker and could even kill a crab or fish that became entangled. Needless to say, those two divers never bothered one again.

There are species which, after rather extensive preparation, are used as food. The product **trepang** is an example of a sea cucumber that is boiled several times and dried to be used in making soup. Most sea cucumbers have the five long muscles that can be stripped off after skinning and fried or used in chowders.

REVIEW QUESTIONS

1. What is the water vascular system, and in general how does it work?
2. What are the pedicellariae used for?
3. Why is the sea star called the Crown of Thorns so important?
4. How does a basket star feed?
5. Only one echinoderm is used in any amount for human food. Which one is it, and what part is eaten?

Figure 20–1 These are examples of rock-boring clams. They bore into rock and spend their entire life trapped in the cavity which they create and then grow into. These have been removed by splitting the rock in half.

Chapter Twenty

MISCELLANEOUS

INVERTEBRATE PHYLA

DEFINITION OF TERMS USED IN CHAPTER 20

Aboral end: The end of the animal away from the mouth.
Casting: A term used for the solid waste left by some animals.
Kelp frond: The leaflike part of the brown algaes.
Lophophores: Hollow, ciliated tentacles found on Phoronida, Brachiopoda, and Bryozoa.
Paedogenesis: A term used when a larval form reproduces without maturing into the adult form.
Polychaeta: Order of marine annelid worms that is widespread and very common.

There are many phyla which, although equally important in many ways as the ones discussed in more detail, are not so significant to the beginning student for a variety of reasons. The main reasons are their relative impact and their visibility and accessibility within the environment, along with the time and space limitations of this book. These animals are *significant;* they are not discussed in detail here because of space.

CTENOPHORA

The ctenophores in many ways look like the jellyfish cnidarans. They are different by not having nematocysts or stinging cells, or a more advanced digestive system. They do have eight plates or rows of cilia running from the aboral toward the oral end. These cilia resembling combs are responsible for the common name, **comb jellies.** The animal generally moves in the direction of the aboral end although most can also move orally if they choose. The group is divided into two classes: those with tentacles (generally two long tentacles extending from pouches near the mouth) and the forms without tentacles. These animals are nearly all pelagic and are part of the zooplankton. They are carnivorous, eating other planktonic forms. Off the West Coast of the United States, the sea gooseberry (*Pleurobrachia*) and the sea vase (*Beroe*) are common forms, while on the East Coast a larger, very luminescent form (*Mnemiopsis*) is common. All ctenophores are hermaphroditic and generally cast their eggs and sperm out through the mouth.

PLATYHELMINTHES

This phylum comprises the group known as **flat worms.** The flat worms include flukes and tapeworms that are parasitic on marine as well as land animals, but are not within the scope of this book. The free-living flat worms of the class Turbellaria are what we find when we turn over rocks. They are small, flat worms with a single oral opening and no anus, as in the cnidarans and comb jellies. Attaching to a large piece of meat, they extrude digestive juices through a long pharynx and out their mouth on to its surface and dissolve away a portion of it, which they then can take into their system. This capability allows them to eat from large chunks of dead meat and not have to catch things smaller than themselves. One order, Acoela, has an algal plant growing in its body and uses the food produced by the plant for itself. This symbiosis is so complete that the worm has lost most of its digestive system from nonuse. The mouth on the flat worms is generally on the underside and in the middle of the body, not on the end as in most other

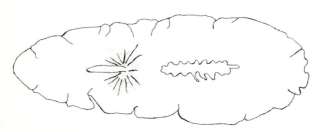

Figure 20–2 Flat worms are very hard to identify. Their taxonomy is based for the most part on internal features.

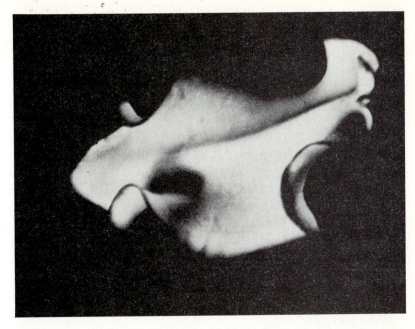

Figure 20-3 Some of the flat worms swim well by rippling the edges of their bodies, as shown here. (Photo by Victoria Von Zedwick)

animals. The large swimming types, which get up to 15 or 18 centimeters in length, are members of the order Polycladida. These polyclad worms are magnificent to watch swimming. Their undulating bodies, together with their bright colors, create a kaleidoscopic effect as they move through the water.

NEMATODA

The **round worms** are considered to be the most numerous multicellular animals in the benthic ecology. They are mostly microscopic, and a large number of them are parasitic. They are slender, smooth worms with a complete digestive tract, that is, a mouth and an anus connected by a straight intestine. Known for their specific parasitism of species, no animal, land or sea, is immune to these parasites. Specific nematode species parasitize specific organs of specific species of host animals and plants. In humans, a few of the well-known ones are pinworms, hookworms, and trichina. They have separate sexes and produce many thousands of eggs. One of the longest types of nematodes reaches 6 feet and can be found in the ocean sunfish (*Mola mola*) living in the flesh, not the intestine. Although undoubtedly of major importance in the general ecological pattern, the nematodes are not studied with the enthusiasm that many other groups are and, consequently, are not as well described and defined as they could be.

NEMERTEA

Nemertea are referred to as **ribbon worms.** Some get quite long, over 3 meters when extended, but are not big around when stretched out. Many have the ability to extend or contract and can change their length as necessity demands. In the Antarctic, fish traps made from small mesh hardware cloth could not keep these animals away from the bait. When the worms approached the wire they were about 30 centimeters long and 3 centimeters in diameter. They would then squeeze through the ¼-inch wire mesh and regain their shape on the inside of the trap. Many times when a trap was pulled, they would be halfway through the wire with a large bulge on either side. The outstanding characteristic that separates them from other forms is their **proboscis.** The proboscis takes many forms, even that of a poisonous dart, but it works basically the same in all species. Connected to a fluid-filled sac, the proboscis is extended when the sac is squeezed by muscle contraction. The length of extension is, in some cases, the entire length of the Nemertea worm. This proboscis is used to capture food.

Some planktonic forms of Nemertea have modified body shapes to swim in the pelagic environment, but most are benthic and need colder water to do well. The Antarctic and Arctic waters abound with them. Although most of this group are free living, there are a few species of parasitic types, generally found in clams. The parasitic types have lost the extendable proboscis and use a sucker disk to attach to their host.

SIPUNCULOIDEA

These fleshy worms, called **peanut worms,** are all benthic and all marine. There are only some 250 species, ranging in size from 0.25 centimeter to as big as 50 centimeters. They live in burrows and are well adapted for that type of habitat. The posterior end of the body is larger than the front section and tends to act as an anchor that cannot be pulled out of the burrow. When disturbed, it inverts its head end back into the body. The anterior end is thus protected. It has a mouth, generally with a ring of tentacles around it. In some species, these tentacles are extensively branched and in a few, hardly present at all. The tentacles are used in feeding. The anus is located about one-third of the body length from the mouth. This is an excellent body plan for an animal that lives in a burrow. The waste is excreted outside the bur-

Figure 20-4 The peanut worm gets its common name from its shape. This one is partially drawn up. When they draw up tightly, they appear even more like a peanut.

row instead of at the bottom of it, where it would necessitate the flushing of the burrow. The most common example of this group is the genus *Sipunculus* spp., found in California, Florida, Europe, and Japan. It ranges from 15 to 25 centimeters in length and is commonly dug out of the sand where it lives. When disturbed, they pull themselves up tightly with the head end inverted into a shape much like that of a peanut. From this resemblance comes their common name, peanut worm.

ECHIUROIDEA

The Echiuroidea resemble, in many ways, the peanut worms discussed above. They differ in having the anus at the opposite end of the animal from the mouth. This necessitates a straight intestine the entire length of the body, whereas the peanut worms have a coiled intestine reaching only one-third the body length. The burrow must be flushed of waste matter. This is accomplished by creating a U-shaped burrow and pulsating the water through the burrow by contractions of the body.

Figure 20-5 Spoon worms live in mud or sand. They build permanent tunnels and are often confused with the peanut worms.

Some of these **spoon worms,** as the Echiuroidea are called, "spin" or, more accurately, secrete a spider web-type net of mucus at the entrance to their burrow. As the water is pumped through the burrow, small particles of food are caught in the net. When the net accumulates food material, the worm sucks it in and digests it. The name "spoon worm" is derived from the spoon-shaped proboscis that is present in most of the group. One type found off Japan has a proboscis over 125 centimeters long when extended, and a body length of only 38 centimeters. One genus has been observed to take in water through the anus into an intestine that is looped back and forth several times in the body. The water is retained until the oxygen is extracted, and then it is released. Most of that group, and probably this one also, take in most of their oxygen through the skin. The U-shaped burrow is also a perfect living place for other small animals, and many take up residence in the burrow with the spoon worm. Some common ones are the pea crab, a goby fish, and even a type of clam, to mention a few. These commensal organisms benefit from the circulating water and protection and do no damage to the worm.

CHAETOGNATHA

The members of this phylum are rather complex worms, rarely growing over 2.5 centimeters in length and spending their life as pelagic planktonic forms. They are transparent and very difficult to see in a water sample unless a microscope is used. They have stiff bristles around the mouth, and the name chaetognath, meaning "hairy jaw," is derived from this feature. These little, transparent worms are ferocious feeders and attack other planktonic forms to such an extent that they must be considered a major predator of the smaller zooplankton. Their rather straight form and speed in the water have given them the common name of **arrow worm.** Even as they are preying on forms smaller than themselves, larger organisms are catching and devouring them. Very common in the plankton, they are a major part of the energy cycle. The arrow worms are also sensitive to even slight variations in ocean water. Because they tend, as a species, to be attracted to or repulsed from various naturally occurring conditions, they are often useful as indicator organisms for certain oceanic conditions.

PHORONIDA

These are worms that generally do not exceed 25 centimeters in length, live in burrows in mud flats as a general rule, and are only about 15 species in number. They have one interesting characteristic, **lophophores.** A lophophore is a ridge with hollow, ciliated tentacles located around the mouth. Two other phyla also have this feature, the Brachiopoda and the Bryozoa.

BRACHIOPODA

These delicate little shelled animals are called **lamp shells.** They are easily mistaken for mollusks because their two shells resemble a common bivalve mollusk shape. The shell of the brachiopod is a dorsoventral (top and bottom) shell, while the bivalve mollusks are lateral (left side, right side) shells. The lamp shells live attached to the substrate either by cement directly on the shell or by a stalk from the end of the shell. Of the some 30,000 species that have been identified, all but approximately 300 are now found only in fossil form. Their fossils are useful to the geologist in making various geological determinations. The lophophore is large and double-spiraled in most forms and is used to catch food.

Figure 20–6 Bryozoans take many shapes, from very delicate colonies like this one to very tough encrusting colonies right in the surfline.

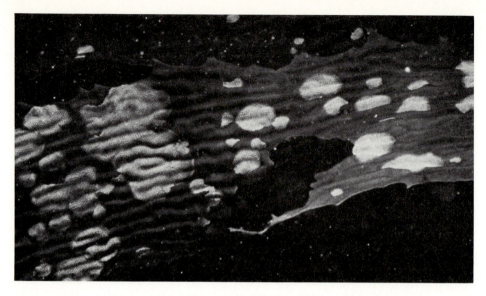

Figure 20-7 The light spots on this blade of kelp are colonies of bryozoans. These small encrusting organisms grow on almost anything.

BRYOZOA

The Bryozoa are the third group that contains lophophores. At present they are much more important than the other two because of their large numbers and encrusting habit. Although they are very common and seen by nearly all beachcombers, they often go unnoticed to the casual observer because of their generally small size.

The division of this phylum into two separate phyla, the Ectoprocta and the Endoprocta, was based mainly on the position of the anus. When it is located outside the lophophore the animal is an Ectoprocta; when the anus is inside the lophophore, it is an Endoprocta. These groups are often called the **moss animals.** They grow on almost anything in a variety of forms ranging from a hard crust only 1 millimeter thick covering a kelp frond to an independent colony that looks like coral if it is hard or algae if it is soft. There are over 4,000 species, all of which have different forms and most of which are encrusting on some other substrate. When examined, any benthic specimen is likely to show some form of ectoproct bryozoan growing on it somewhere. The endoprocts are far fewer in number of types and not found so often. One has to be a knowledgeable biologist to tell the two types apart; the internal structural differences, although definite, are difficult to discern.

Most members of the entire group of bryozoans are colonial; therefore, large numbers of them are generally found at any one time. Each possesses its own digestive track, a U-shaped tube, and it is not connected with the other, as are the Hydrozoa. Bryozoans are truly separate animals that live in a colony much the same as corals. They even are mistaken for corals at times because of the similarities between the two groups. All of the bryozoans are filter feeders, having small tentacles covered with cilia and mucus around the mouth. The cilia set up a current of water, beating and moving microscopic particles into the mucus net where they are trapped. The bryozoans need a substrate to grow on and will try almost anything: rocks, kelp, shells of crabs, clams, brachiopods, pilings, junk that has been towed into the sea, and even the body of a sipunculid worm. Because they are able to reproduce by budding, a colony can get a quick start. By budding, a few individuals can produce many more that produce still more, and the colony spreads out rather rapidly. They also reproduce sexually, and it is these sexually produced larva that join the zooplankton for a while and then settle on something to start a new colony through the budding process. These are entertaining animals to work with in the classroom. They are small enough that without a microscope you really can't see much, but large enough that with an average microscope you can see almost everything. Bryozoans are easily collected almost anywhere there is salt water.

ROTIFERA

The rotifers are not a particularly important marine form. They are mostly fresh water dwellers. They are interesting, however, because their body shape is much like a trochophore larva. The trochophore larval stage is one of the stages that both the mollusks and the annelids pass through on their journey from egg to adult. The presence of a common stage in the development of two different animal types is generally taken as evidence that both types had a common ancestor. The rotifers, then, may share this common ancestor with the annelids and mollusks, or the rotifer could represent a larval form that adjusted to a different environment by not maturing to an adult form. To do so the larva would need to become sexually mature while still in the trochophore stage. Although this is just speculation, it does appear that there was a common ancestor for the three forms.

ANNELIDA

The annelids are well-known to everyone because the common earthworm and fishing worm are annelids. The main visual characteristics of the group are the rings around the body that give the worm a segmented appearance;

consequently, they are called the **segmented worms.** These worms are, in fact, segmented internally in such a way that many of the organs are repeated in each segment. We are only concerned with the two main groups of marine annelids. Class Hirudinea, the leeches, are rare enough that we need mention only that they have a very wide range of occurrence, but are not in large numbers anywhere. They are fairly common on sharks and rays. The other group, the class **Polychaeta,** includes the majority of segmented marine worms—over 5,000 species.

Representatives of the Polychaeta are found in all environments. Polychaeta have a trochophore larva that swims in the plankton and is sometimes mistaken for a rotifer, as mentioned above. The family Tomopteridae is a totally planktonic Polychaeta, but the majority of Polychaeta are benthic. They inhabit a shallow as well as deep sediments and can nearly always be found if a handful of marine algae is pulled loose and the holdfast inspected. They are found as burrowers in sand and mud, free swimmers and crawlers, and living in tubes which they build out of a variety of materials. The group is large enough for many special adaptations to have been made within it, such as nest building with sea grass, hard calcium tubes in which to live, leathery mucous tubes, delicate sand tubes, filter feeding nests, and strong jaws to catch prey. This is such a varied class that almost any modification could be found if one looks hard enough for it.

One of the most striking modifications is found on the members of this group we commonly call feather-dusters. The species thus called live in a tube that they create. The tube is generally calcareous and many times has

Figure 20–8 The many species of annelid worms are found in all parts of the oceans. The characteristic segmentation is easily seen.

Figure 20–9   Colonies of annelid worms often resemble beehive honey-combs. These create the tube by sticking sand particles together with mucus, and may cover a hundred or more square feet.

a sharp point at the entrance to keep other animals away while the occupant is withdrawn. When the feather-dusters are out and feeding, they have a gill network combined with cilia that looks like one or two old-fashioned feather dusters sticking out of the end of the tube. They withdraw back into the tube very quickly at the first sign of trouble. Some are light-sensitive and will withdraw if a shadow passes over them. For whatever reason, when they withdraw, they do it so rapidly that it is generally impossible to watch. One second they are there; the next part of a second they are gone. They are a favorite subject for underwater photographers because of their delicate structure and varied colors, and they are a challenge to photograph before they withdraw.

The feather-dusters are plankton feeders and use the gill duster to filter the water. Others use their tubes as a water channel to guide water across their body or through a mass of mucus to catch various detritus that they use as food. The third main feeding pattern is as predator. These predaceous forms have teeth, sometimes on a proboscis that can be extended a short way to catch other organisms. They all have hooklike jaws to hold the prey, and most of them will eat anything they can catch. The majority of the predaceous annelids do not burrow, but rather live in the area of protected cracks, seaweed, and rocks. Many of those that do burrow seem to share their burrows with other guests, most commonly with various small crabs, shrimps, and other worms. In some cases, the annelid is the visitor and lives on or with some host of its own, such as a sea cucumber or sea star.

The spawning habits are as highly varied as the individuals in the group.

Figure 20-10 The Christmas tree worm is a beautiful annelid that can retract into its tube and disappear in a fraction of a second. It feeds by fanning out and catching plankton.

One particular annelid is famous for its breeding pattern. The palolo worm of the South Pacific is so specific about the month, time of moon, and level of tide, that its breeding is predicted in a manner similar to the way we predict a grunion run in Southern California. The palolo worm is a fine treat and is gathered for food by the locals of Samoa and Fiji. They are easy to gather during the nights they spawn, because they come to the surface by the thousands to swim in large masses. This assures that there will be a sufficient concentration of sperm in the water to ensure fertilization. The surface swimming worm is not the entire worm, but just the reproductive portion that was detached from the main body to rise to the surface. The annelids are truly interesting creatures, for they include so many life styles, body shapes, and interspecific relationships that we still need much more research to truly understand the phylum.

ENTEROPNEUSTA (HEMICHORDATA)

This small phylum contains less than a hundred species. They are mud dwellers for the most part and are seldom seen out of their burrow. There are two genera common to both the eastern and western coasts of North America, *Balanoglossus* and *Saccoglossus*. Because of the shape of their head, they are called **acorn tongue worms** and **bag tongue worms** (*balano,* "acorn"; *sacco,* "bag"; *glossus,* "tongue"). They are not commonly seen because of their burrowing habit and their delicate nature that makes them difficult to dig out without breaking. The majority of the group are bright colored; some are orange, red, yellow, or combinations of colors. They have what looks like a ring around the "neck," and the proboscis, which looks like the head, is shaped like an acorn. When they burrow, they take in mud through the mouth and eat their way through the mud or sand, with the ring around the neck setting the diameter of the burrow. Slime on the body creates a casing for the burrow by sticking sand or mud particles together. The mud that is taken in the mouth has the water squeezed out over the gills for respiration and the organic matter digested out for food. The remainder is pushed out of the burrow as a casting. As far as we know, these are mainly shallow water forms, with only a few species being from deep water of 5,000 meters or more.

CHORDATA: CLASS UROCHORDATA

The **tunicates** are an interesting group for several reasons. In the larval form, they contain a notochord that indicates a direct relationship with higher forms. Because of this notochord they have been classified as a class of phylum Chordata for years. Of late, some researchers have separated tunicates into their own phylum. Others consider them to be a subphylum. They are a difficult group to define.

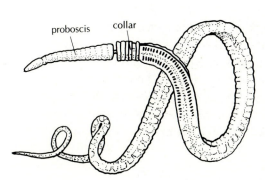

proboscis   collar

Figure 20–11 The acorn worm is a free-living creature of the inter- and subtidal zones.

In their tunic or outside "skin," they have a cellulose-type material that is extremely rare in the animal kingdom but is common in the general body material of most plants. They are quite common and are found growing in most intertidal regions from pole to pole, being brought up in deep dredges as well. These three features, as well as others, make them a group of particular fascination for the biologist.

The higher forms of animal life, the vertebrates, start in larval form with the development of a notochord which, as they change to adult form, develops into the vertebral column we call the backbone. In the tunicates the notochord is present in the larval form, but it never develops and either becomes rudimentary in the adult or is lost altogether. The larval tunicate generally appears as a swimming, tadpole-shaped creature with the notochord running down the length of the tail. The common form of attached or sessile tunicate has such a larva, and when it is ready to attach to some substrate, the adhesive glands on the anterior end attach by way of a sticky mucus "glue." The mouth is located near the point of attachment at first, but it is moved away from it by the rapid growth of the area between the mouth and the point of attachment. Most of the tail is dropped off, and the rest of it is absorbed into the body. The result is an attached animal with the mouth seaward from the substrate and no notochord left in the animal. The tunic grows up from the base and covers the animal with a tough, leatherlike skin composed of tunicin, a cellulose material that gives rise to the name "tunicate."

The Ascidiacea (sessile) tunicates are often called "sea squirts." When exposed at low tide they have the habit of contracting their saclike body and squirting out a stream of water. If they are disturbed, they will almost always squirt. The adults have two openings in the tunic: the incurrent opening and the excurrent opening. They can squirt out of either or both. The ascidians may be solitary (simple) or grouped together and connected by a common tunic (compound). In the compound tunicates, it is not uncommon for each individual to have its own incurrent siphon and expel the water into an excurrent cavity for the entire colony. The colonial varieties reproduce asexually by budding to create the colony, but they can also reproduce sexually with free swimming larva to create new colonies.

There are two other classes of tunicates, both of which live in the open sea as planktonic forms when adults. The class Larvacea gets its name from the fact that it never develops beyond the larval tadpole form. It reproduces sexually directly from the larva shape, a biological phenomenon called **pedogenesis.** It is hermaphroditic, with the sperm maturing first and the eggs later, so it can't fertilize its own eggs. Larvacea are very common planktonic forms with a mucous covering over the head to entrap other planktonic organisms used as food. Class Thaliacea is also a pelagic group. They are different from the ascidians in having a clear tunic, and they are transparent.

In addition, the incurrent and excurrent siphons are at opposite ends of the animal. The Thaliacea are often called **salps.** In the first part of their life cycle they are colonial and occur in aggregate forms, meaning that they are found in chains of several hundred, breaking loose when they mature. One order, Pyrosomida, is colonial during its whole life cycle and shares a common excurrent pore, while taking the shape of a tube up to 60 centimeters long. The highly luminescent genus Pyrosoma is a widely distributed example of this type of salp. The reproductive process in this class is so varied that a detailed discussion would take many pages. The reproductive process varies from sperm and egg loose in the water to a single egg being kept attached to the parent by a structure similar to the mammal's placenta. In some cases, they also show an alternation of generations, as many plants do, with a sexual generation followed by an asexual generation and reverting to a sexual generation again.

CHORDATA: CLASS CEPHALOCHORDATA

The members of this group are small fishlike creatures called *lancelets.* Seldom seen, they are found in shallow sandy areas and in burrows beneath the sand. Because of their form and structure, they are considered to be of ancestral importance to the higher vertebrates and are considered by many researchers to be different enough to be classified as their own subphylum. They have a notochord, which is the main supportive structure of the adult form, and a dorsal nerve cord above it, as is found in higher groups of animals. They also have gill slits like those found either as functional or vestigial in higher animals. Because of pollution they are becoming much harder to find in recent years. Their habitat is normally in quiet bay areas, and because humans have taken to dumping various pollutants into the waters that empty into most of the bays of the world, these pollutants filter into the sand, and being very sensitive to its environment and having narrow tolerance levels, the lancelet cannot survive.

REVIEW QUESTIONS

1. One phylum is particularly well-known for the fact it is a parasite. Which one is it?
2. The arrow worm is often used as an indicator organism. Where is it found?
3. Which phylum encrusts other organisms and often goes unnoticed?
4. Most marine Annelida belong to what class?
5. What is most unusual about the skin of the tunicates?

Chapter Twenty-One

CHORDATA: FISH

DEFINITION OF TERMS USED IN CHAPTER 22

Anadromous: Fish that lay eggs in fresh water and mature in salt water, such as the salmon.
Chondrichthyes: Class containing sharks and rays.
Agnatha: Class containing lampreys and hagfish.
Lateral line: A special sense organ which picks up and locates low frequency vibrations such as a fish swimming; well developed in most pelagic sharks.
Osteichthyes: Class containing the fish that have bones.
Spiracle: Hole in the top of the head of the sharks and rays, used to take in water for respiration in bottom dwellers.
Swim bladder: A gas-filled organ found in most bony fish, but lacking in sharks and rays, that acts as a buoyancy control device to keep the animal neutrally buoyant in the water so it will not sink or float.

The phylum Chordata is of such interest that many books have been written on each individual class and many books on just a few species within a single

Figure 21-1   The moray eel is a common shallow water resident.

class. We will introduce the main groups so the students may recognize them as separate classes and develop an appreciation for some of the differences and adaptations found in each class. As in the other phyla, there is a choice of ways to classify these animals. We have chosen what seems (at least to me) the least complicated. The phylum is divided here into classes. This chapter will be about those classes that live only in water: Agnatha (lampreys and hagfish), Chondrichthyes (sharks and rays), and the Osteichthyes (fish with true bone). The chordata classes that have only a few marine representatives are discussed in the next chapter. These are the reptilia (sea snakes and sea turtes only); aves (sea and shore birds only); and mammals (marine forms only). Because each of these groups is so well discussed in specific books that are readily available from any library, we need here only to introduce each one.

CLASS AGNATHA: ORDER CYCLOSTOMATA

The Cyclostomata are characterized by being long and slender (eel shaped) and can be recognized quickly because they do not have any jaws. The two main types in this class are **lampreys** and **hagfish.** The lampreys are found in fresh water lakes and streams as well as in salt water and have become a major problem in the Great Lakes area, where they feed on and kill the fish. Hagfish are found only in salt water and feed mainly on dead fish.

The lamprey has been the subject of a great deal of study because of its damaging effect on fish in the Great Lakes. Many methods have been tried to be rid of it. The species found in the Great Lakes region are parasitic and attach to the fish by using their mouth as a sucker. They have many small teeth, located in rings around the mouth, and are able to rotate them back and forth to bore a hole in the side of the fish. They then suck the blood out of the fish until it dies or is so weak that it dies after the lamprey has left it. The adult lamprey swims upstream and lays eggs in the gravel beds of the streams. The larvae are unlike the adult and spend several years in the stream, where they do little damage. When they mature, they swim back to the ocean. The problem in the Great Lakes is that the lampreys have become at home in the lakes. Instead of going to the ocean they stay in the lakes and feed on the limited population of fish that live there.

The hagfish live in deeper water, generally below 100 meters, and feed on dead material. They also have a circular mouth, but soft sensory tentacles instead of the teeth the lampreys have. Hagfish are able to produce mucus or slime from their bodies and are very hard to handle because of it. Also

Figure 21-2 The hammerhead shark's eyes are located at the ends of a broad extension from the head.

called slime eels, they fit the name well. Their reproduction is interesting, as they differ from the other vertebrates by being functionally hermaphroditic. They only produce sperm or eggs at any one season, but are capable of producing sperm one year and eggs the next.

CLASS CHONDRICHTHYES

The Chondrichthyes include the sharks, rays, and chimaeras. Although these three generally look quite different from each other, they have many features in common. None of them have any true bone; instead, their skeleton is composed of cartilage. They have strong jaws, and the mouth is located on the underside of the body, while the eyes are on top. They cannot see anything as it enters their mouth. Some sharks solve the problem of locating where to place the mouth by touching their food with their nose first to get a reference point before placing the mouth on it. The time between the nose touch and the bite need be only a fraction of a second.

The entire class has tough skin, most with a noticeably sandpaperlike texture. This raspy skin is the result of having **placoid scales,** which are like small teeth, present in the skin. Some of the larger sharks have such rough skin that the skin is dried and used as sandpaper. The males and females are easy to tell apart in this group. The males have a set of **claspers** extending

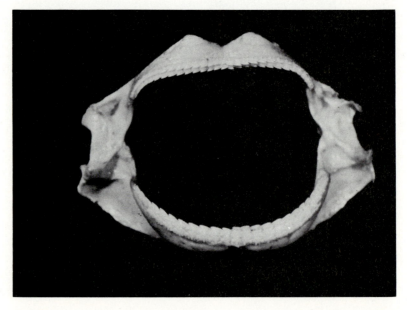

Figure 21–3 Sharks have several rows of teeth. As a tooth breaks off, a new one moves up to replace it.

Figure 21-4 The shape of a shark's tooth is characteristic of its species. The shark can be identified even if one tooth is found.

from the pelvic fin area near the cloaca. These claspers are shaped like two fingers and the sperm is conducted by them from the male into the cloaca of the female. Fertilization is generally internal. The females of some species lay eggs, a few of which are strangely shaped and are often found on the beach; other females retain their eggs in the body until they hatch. The sharks retained during development are not nourished by a connection to their mother like a mammal through a placenta. The eggs are merely retained by the mother in a portion of her oviduct for protection while they develop. Oxygen generally is transferred between the mother and the developing embryo, but no nourishment.

One of the most easily recognized features of this group is the external gill slits. There are generally five pairs of gills, with a few forms having six or seven and the chimaeras having only three. This number serves to differentiate them from the bony fish, which only have one external gill opening. The chimaeras have a membrane covering the gill slits, and may confuse the layperson because they appear on first glance to only have one gill slit like the bony fish.

The sharks are well-known because of the reputation given them by writers who exploit sensationalism to sell their books or movies. These stories have created an unrealistic fear of sharks. Out of the hundreds of species of shark, in reality only a handful are dangerous to humans, and these are responsible for only a few shark attacks a year among the millions of people who are exposed to them through swimming and diving. Dogs kill more people each year than do sharks. Fictional books about sharks are fun to read, but they should not be allowed to affect the rationality of the reader.

Sharks are graceful swimmers and beautiful to watch underwater. The

upper portion of their tail fin is longer than the lower portion, a good characteristic for identification of the group. The one exception to the uneven tail
is the white shark, which has an even tail fin like most bony fish. It also
maintains an internal temperature a few degrees higher than the water in
which it lives, which makes it, in a sense, warm blooded. Because of its
higher metabolism, it feeds more often than other sharks and has the de-
served reputation of trying to eat almost anything it encounters. The white
shark is a rare fish, but it has a wide range and has been reported in most
oceans at one time or another. It is most common off Australia and in the
Indian Ocean.

Many people believe that sharks have to keep swimming to breathe.
This is not true. It is true, however, that they must keep swimming or they
sink. They do not have a swim bladder like the bony fish to act as a buoyancy compensator, so their only means of keeping off the bottom is to swim.
This condition has led to two distinct groups, the bottom dwellers and the
open water forms, or the benthic and pelagic forms. The pelagic forms have
adapted to the constant movement through the water and do rely a great
deal on the water passing in the mouth and out over the gills for respiration.
Of the forms that lie on the bottom, many bury themselves under the sand
or mud and take in water through a pair of holes in the top of the head
called *spiracles.* The water passes in the spiracle and out over the gills. This
is accomplished by a bellows action of closing the gills and expanding the
throat area, then closing the spiracle and contracting the throat area to push
the water out over the gills. Because the spiracle is directly on the top of
the head, many forms can lie completely buried in a soft substrate and be
effectively undetectable. The rays have adapted to doing just that. They are
like sharks in most ways except body shape, for the ray's body has become
flattened. Many fish are flattened, such as halibuts and flounders, but most
bony fish are flattened laterally, side to side. Rays are flattened top to bottom, or dorsoventrally. This flat body is the perfect shape for hiding on the
bottom and for digging up crabs and shelled animals to feed. Sharks and
rays have an intestine different from the bony fish, in that it is quite large in
diameter and short. It contains a **spiral membrane** that makes the food go
around and around as it passes through, delaying it long enough to be digested and absorbed. They also have an extremely large liver, which aids in
digestion by producing bile and stores a great deal of the digested nutrients.

Like any large group, the Chondrichthyes have developed many
unique modifications. The rays have several means of protection. Some
have developed a large toothlike stinger located on their tail that can be
whipped up and used to protect their dorsal side while they are buried in
the sand. Another ray has developed a type of battery and can deliver a
sizable electric shock if touched.

The pelagic sharks have developed an extremely efficient organ for locating food called the **lateral line.** The lateral line runs from the head of the shark to the tail along the dorsal side. It is paired with one on each side of the shark. Just as we see two images, one with each eye, and our brain triangulates to give us the distance to the object, the lateral line triangulates to give the distance of a prey to the shark. The difference is that our eyes are only an inch or two apart, while the lateral line on a shark could well be many feet long, giving a much more accurate location. The shark can launch an accurate attack on its prey even in total darkness. The lateral line is sensitive to low-frequency vibrations such as those sent out by the tail of a swimming fish. If the fish is injured and struggling, the shark can recognize this condition by the pattern of vibrations sent out by its movements and will normally respond by coming after the struggling animal. Any predator responds quickly to a wounded prey because it is easier to catch. The lateral line system is found on most kinds of fish, but the pelagic sharks have one of the best developed systems. The lateral line is noticeable because it has pores in the skin or, in bony fish, special tube scales running along its length. These pores and tube scales let water into a groove in the skin that has sensory organs. The lateral line sensory organs are found developing on the surface of the skin in most embryonic fish; as the fish matures, a groove differentiates along the side of the fish, and the sense organs sink down and

Figure 21–5 Fishing is one of the important uses of the sea. The millions of dollars spent on sport fishing is a large portion of the tourist economy.

are covered over, being exposed only through the pores in its covering scales or through tubes to the surface. On the more primitive forms, such as the chimaeras and the frilled shark, the sense organs are located in open grooves. The system seems to combine the qualities of hearing and touch. Humans have no sense that compares to it.

The rays generally are found as benthic types, and sharks are generally considered to be mostly pelagic. Exceptions, especially with bottom-dwelling sharks, occur in most areas. The largest form of ray, the manta reaching 8 meters across, and the two largest sharks, the basking and whale sharks, both reaching around 15 meters in length, are plankton feeders. By moving down the food or energy chain to the bottom layer, more food is available more often, and it takes less energy to harvest it. The largest whale, the blue whale, is also a plankton feeder. Perhaps some day through neces-

Figure 21-6 The puffer fish, one of the more interesting species, fills itself with water and extends sharp spines that make it impossible for a large fish to swallow. It is commonly called a porcupine fish.

Figure 21-7 The kelp fish stays close to the giant kelp plant for protection. Because its body design and coloring are much like the kelp, it becomes hard to see.

sity humans will become skilled enough to harvest this vast source of food that these giants of the seas have been harvesting for thousands of years.

CLASS OSTEICHTHYES

The bony fish, as they are called because of the bone material that comprises their skeletons, include most of the fish with that we are familiar: bass, cod, perch, tuna, halibut, sardines, etc. Besides their bony skeleton, they have other easily recognizable characteristics; a **single gill slit** on each side, the mouth generally at the front of the body (not on the underside), and a tail fin with the top and bottom portions nearly the same size. Internally, the intestine of the bony fish is a long, slender, coiled tube similar to that of humans, instead of a short, large one with a spiral valve like the shark's. Both types of intestine give sufficient surface area to digest the food and absorb it. Unlike the sharks and rays, the bony fish have external fertilization of the eggs. The female generally produces large numbers of eggs and deposits them into the water. The male deposits sperm, normally called *milt*, into the water. The milt disperses and fertilizes the egg on contact. Fish of different types have different breeding patterns. Some school together during the breeding season, so the eggs and milt are dispersed in the same area. A few actually pair off, and still fewer are capable of self-fertilization. Although this may seem like a haphazard way of fertilization, it is, in fact, quite efficient. The plaice, a type of fish, produces between 250,000 and 500,000 eggs during a single season. When samples are collected, very few unfertilized eggs are found. Some fish, such as salmon and lampreys, are called **anadromous,** which means they leave the ocean and deposit their eggs in fresh water streams. For many species, the trip up the rivers, the respiratory changes encountered in fresh water, and the general effort needed to make the journey and breed are more than they can stand, and they die after breeding. Other forms, like the fresh water eel of Europe, spend their adult life in fresh water and migrate to the sea to breed, just the opposite of the salmon. In New Zealand, one of the delicacies used as food is the two-inch larval form of such a fish called white bait. It is caught in hand-held nets as it comes up the rivers. The people gather at the beaches where rivers run into the sea to catch the white bait, much as the people in Southern California gather at their beaches to catch the grunion.

The bony fish, because their reproductive methods produce thousands of eggs, have had an excellent chance to undergo adaptations and make specific changes over thousands of years to become well adapted to many specific ecological habitats. A few types of specific adaptations are described below to illustrate the great variance existing in this group.

The flat fish, such as sole, halibut, and flounder, are hatched looking like "normal" fish. As they develop, they undergo changes that include the

Figure 21–8 A small royal gramma, or fairy basslet, hangs upside down under a rock. Many of the smaller fish that live in reefs orient themselves to the nearest part of the reef instead of the gravitational pull of the earth.

Figure 21-9 The diver is feeding some small reef fish a piece of bread.
In the presence of food, the fish overcome their fear of man.

migration of one eye to the other side of the body so that both eyes are on the same side. The fish then orients itself so the eyes are on top and appears to be flat like a ray. The difference is that the "top" is really a side, and the bottom is really the other side.

In sea horses and pipefish the male has a brood pouch and the female lays her eggs in it. The eggs are cared for by the male and when they hatch, the young swim off.

The remora is a fish that has developed a sucking plate on the top of its head. Attaching to larger fish for a free ride, it then gets the scraps that are left after the larger fish eats. Remoras are most often associated with large sharks. Once, while diving, the author had a remora try to attach to his leg. It was difficult to chase off.

Mola mola, the ocean sunfish, is a poor swimming fish that attains over 2,000 pounds in weight. Shaped like a perch that has had the tail section cut off, it feeds mainly on jellyfish and drifts with the current. It is said to be the largest form of zooplankton. Because of its slow speed through the water (about 3 miles per hour maximum) and its large size among the plankton, it has become the home, or host, for an extremely large number of parasitic creatures. It is often taken by biologists who are studying parasites, but the flesh is not good to eat.

Grunions are famous along the Southern California coast because they come ashore to breed. During the very highest spring tides between April

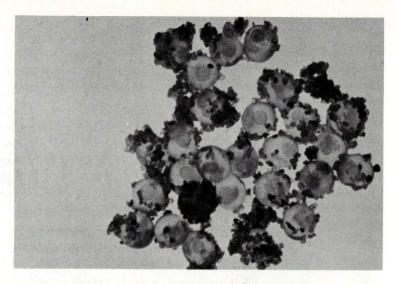

Figure 21-10 The female placed these grunion eggs under the sand on a high spring tide to hatch. They were removed by students after the tide had gone out and brought back to the laboratory for study.

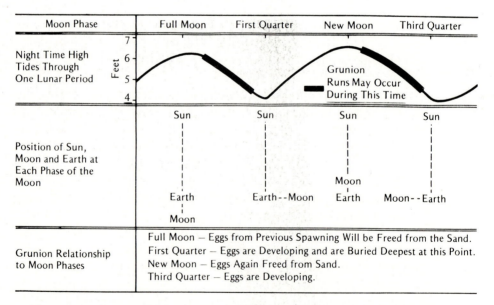

Moon Phase	Full Moon	First Quarter	New Moon	Third Quarter
Night Time High Tides Through One Lunar Period				
Position of Sun, Moon and Earth at Each Phase of the Moon				
Grunion Relationship to Moon Phases	Full Moon — Eggs from Previous Spawning Will be Freed from the Sand. First Quarter — Eggs are Developing and are Buried Deepest at this Point. New Moon — Eggs Again Freed from Sand. Third Quarter — Eggs are Developing.			

The grunion is one of the interesting fish found off the coast of California and Baja California. It lays its eggs on the sand during high tides where they develop out of the water. The eggs hatch at the next high tide and larval grunion return to the sea.

Figure 21-11 The relationship of grunion behavior to the sun, moon, and earth tides.

and August, the fish wash up on the beach. The females dig with their tails until only their head is exposed and lay their eggs. The males wrap their bodies around the females and deposit the milt. The milt, a creamy fluid, runs down the female's body and fertilizes the eggs. Both male and female then wash back to sea with the next breaker. The eggs develop, buried in the sand, until two weeks later when they are washed out to sea by the next spring tide.

Some deep water fishes that live where there is no light have a way of producing cold light through bacterial action or the use of an enzyme. One such species, which has a cold light source under each eye, can also shut off the light with a movable membrane. It can turn its "head lights" on or off whenever it wishes.

Many different species of jellyfish, including the colonial hydrozoan called the Portuguese man-of-war, have specific small fish living under their mantles for protection. Even the sea cucumber is not free of a free rider, for there is a small fish that moves into the anus of the sea cucumbers tail-first to stay protected during the day. At night the fish emerges to feed, then goes back for protection the next day.

A fish found in the Indian Ocean has a colony of hydroids growing on

Figure 21–12   This tide pool sculpin, a typical bony fish, shows why sculpin are difficult to see in the tide pools or while diving. They blend into their surroundings so well that the observer is generally not aware of their presence unless they move. Many are very small and go entirely unnoticed.

it. The hydroids protect the fish by making it look like a rock covered with growth; and when the fish swims it moves the hydroids through the water so they can feed more easily. This is a true **symbiotic** relationship.

Some fresh water fish also have unique adaptations, such as the ability to breath air and stay out of water for extended periods of time, climb trees, and shoot insects off of bushes by squirting water at them. The Osteichthyes are truly a remarkable group of animals; if one were to investigate thoroughly, it is likely that almost any imaginable modification to life style or body form could be found.

REVIEW QUESTIONS

1. What are the differences between hagfish and lampreys?
2. How can you tell a male from a female shark?
3. How does the lateral line function?
4. How does the fertilization of eggs differ between the sharks and the bony fish?
5. Explain how the biologists generally use *Mola mola,* the ocean sunfish.

Chapter Twenty-Two

CHORDATA:

REPTILES — BIRDS —

MAMMALS

DEFINITION OF TERMS USED IN CHAPTER 22

Baleen whales: Whales that are filter feeders.
Carnivora: Order containing the seals, sea lions, etc.
Cetacea: Order containing the whales and porpoises.
Chelonidae: Family of the sea turtles.
Hydrophiidae: Family of the sea snakes.
Larus: Genus name for gulls.
Pack ice: Sea ice formed into a mass by the crushing together of frozen sections which have frozen on the sea surface.
Sirenia: Order that is the least known of the sea mammals, containing the manatees, sea cows, and dugongs.
Squamata: Order of the sea snakes.

The animals in this section are parts of classes that are mainly land forms. The species that have taken to the marine environment are the oddities in the class, not representative of the normal life style of the class. Every class has at least a few species that have left the competition of their dry environment to forage in the sea where food is more abundant and the temperature more stable. All of these animals have one thing in common: They did not develop in the sea. They are true land forms that have over thousands of generations adapted to life in a water environment.

CLASS REPTILIA

Normally we assume, without thinking, that a population of animals will grow as time passes. If we stop to think, it becomes obvious that cannot happen over the long term of thousands of years. As one population builds up, generally some other population must decrease to make room and leave food for the new, better adapted population. So it was with reptiles. The reptiles developed as a new group of land animals from the amphibians. They were extremely well adapted to the new land environment and proliferated widely over all but the very cold parts of the earth. For 150 million years they were the dominant type of animal on the land. When mammals started to develop, they soon out-competed the reptiles, and the few species that survive today are all that is left of the fascinating Age of Reptiles. Of those that have survived to be part of our world, three types have marine species. The snakes, turtles, and lizards have members of their clans that have changed form slightly to live in salt water. Crocodiles also have made the adaptation, but not so well as the other three, as they prefer brackish water to sea water. The sea turtles of the family Chelonidae, the sea snakes of the family Hydrophiidae, and the marine iguana of the family Iguanidae comprise the majority of reptiles that have left land to reenter the sea from which their ancestral forms, the amphibians, came.

The Sea Turtles, Chelonidae

Turtles have had the same general form or body shape for well over 100 million years, as we know through our fossil records. The hard shell protects them from their environment very well and has helped in their survival. The land forms are heavy and slow moving because of gravity and the movement limitations placed on them by the hard, heavy shell; in the sea these limitations are much reduced.

The water buoyancy reduces the shell weight, and the animal no longer has to hold the body up off the ground, but rather can move through the environment instead of on it. The feet have become modified into broad paddles and have elongated a bit to give the sea turtles great mobility and remarkable speed underwater. They can swim as fast as many fish. Although they are truly marine animals, they still must breathe air and lay their eggs on land. Since the development of humans on earth, these two traits are becoming the downfall of sea turtles. When they rise to the surface to breathe, humans spear them for food. When they lay their eggs on the beaches, humans gather them for food. Only recently have people realized that the turtles could join their ancestors in extinction if they are not protected from their new predator, humans. Laws have been passed to protect turtles, and the indications are that this time humans may have acted soon enough to do some good and save the turtles.

The sea turtles have made several unique specific adaptations for living in the sea. The turtles normally cannot expand their lungs to any great extent because of the hard shell. This limited air would limit their time underwater. To compensate, the bottom portion of the sea turtle is hinged in such a manner that it does allow for large lung expansion and lets the turtle take in an increased volume of air for a longer stay underwater. To even further increase the underwater time, the cloaca, or anal chamber, has become highly vascular with many blood vessels; and the turtle, by working water in and out through a bellows action, can give off CO_2 and take on O_2 in sufficient amounts to stay under for hours if at rest.

The **green turtle,** the most widespread variety, is found in shallow, inshore waters in most warm seas of the world. It reaches around 500 pounds and is the one most often taken as food. It is protected now in most areas, as are the others.

Figure 22-1 The green turtle is prized for food but is now a protected species in most areas. Commercial turtle farms, such as the one on Grand Cayman Island, raise turtles like a farmer raises steers. They sell them to restaurants.

Not so large as the green turtle, the **hawksbill turtle** reaches only around 200 pounds and does not have quite so large a territory or range. It is also a coastal turtle, found mainly in the Caribbean, the Gulf of Mexico, and the Indo-Pacific, with a few along the coast of Baja California. It has been observed to eat the Portuguese man-of-war and protects its eyes from the stinging cells by keeping them closed. Normally it eats plants and whatever animals it can find or catch.

The **loggerhead turtle** can attain a weight of 900 pounds and is found more often in colder waters than any of the rest. Most common in warm seas, it is also recorded in Nova Scotia and Southern California. Unlike the two above, it feeds mainly on meat. It forages for food over the continental shelf and lives mostly on invertebrates. Although not an open sea animal, it is more often found farther from shore than are the green and hawksbill turtles.

The **ridley turtle** is the smallest and least known of the sea turtles. It is only about two feet long, and although found all along the eastern coast of the United States and western coast of South America and Mexico, it is not normally found in the Caribbean. It seems to eat plants and animals and lives near shore in swamp areas.

Leatherback turtles are perhaps the rarest of the group and also the most adapted to their marine life. They swim faster and grow much larger (well over 1,000 pounds) than any of the rest. This giant turtle is truly an

Figure 22-2   The skua is a type of large carnivorous bird found in the Antarctic region and the North Atlantic. There are several species.

Figure 22–3 The author visiting with a group of Adélie penguins in the Antarctic. (Photo by John Dearborn)

open-ocean form and wanders most of the warm and temperate seas. It has a leatherlike shell, so it is much more flexible than the others. They are not seen often and generally are spotted by boaters, not scientists, so information on them is scarce. It is known, however, that they can be quite vocal when disturbed and will often scream at their molester.

The Sea Snakes (Order Squamata)

A number of species are all related to the cobra-type land snake. Their venom is extremely toxic, but they are fish feeders and do not attack humans. Scuba divers often see them when diving within their range. They are found from Baja California south to Equador and across the Pacific to Africa. The species we have off Baja California, near the United States, is called the **yellow-bellied** sea snake. Sea snakes live entirely in the sea, and only one species comes ashore to lay eggs. They are seen often at the surface taking in air and at times congregate in large schools, miles long, presumably for breeding.

Marine Iguanas

Marine iguanas are found only on the Galápagos Islands. They live on land but feed in the sea, mainly on algae. They are good swimmers and can stay underwater for many minutes.

CLASS AVES, THE BIRDS

The large variety of birds and the difference in the avifauna from area to area make it impractical to elaborate on the group here. Only a few examples of their modifications to the sea are described below. Many identification guide books are specific for a given area. Such a guidebook is recommended for identifying your local species or for those of areas where you travel.

All of the marine birds use the sea to feed and the land to breed. Perhaps the best adapted birds to this type of life are the **penguins.** The penguin has been so successful at its marine life that it has lost the ability to fly. Its wings have become paddlelike so it can swim better. A fine swimmer, it can be seen in Antarctic waters jumping out of the water like a porpoise. During

Figure 22-4 Among the largest of all shore birds, pelicans breed in remote areas along the coast and on offshore islands and are very protective of their nest. They can reach a wing span up to nine feet.

the International Geophysical Year, the author was in the Ross Sea in the Antarctic and one of the few pleasurable pastimes, when we were not working, was to watch the penguins. One member of the expedition moved into a penguin colony where the birds were nesting and lived with them for 18 months. The information he gained on their life style and basic ecology was a great help in understanding these water birds.

The penguin uses its wings to fly underwater. While most birds swim with their feet and steer with their wings, the penguin uses its wings, which have developed special muscles, to swim in much the same way it would fly in the air. It uses its feet for steering. The Adélie penguin returns to its same nesting site or rookery each year. If the sea ice remains frozen from the previous winter, the penguin must then walk to the nest over the ice. Penguins have been known to do this for a distance of over 32 kilometers. The large emperor penguin nests right on the ice and keeps it eggs warm by holding them on top of its feet and covered with down. Even though the temperature may drop to −45°C, the eggs still survive.

Gulls are the best known of the marine birds. They are the scavengers that keep our beaches clean and eat any dead or wounded fish they can find or any live ones they can catch. In most areas the gulls are fairly tame and will let humans come within a few feet of them. One of the pleasant pastimes on the beach is watching this bird looking for food. Gulls have a worldwide range. The genus *Larus* has many species; one or more of them can be found almost anywhere one travels, even in lakes miles from the sea.

Figure 22-5 Gulls, the most common shore birds, nest on offshore islands or any area of solitude.

In Utah the gulls are protected and have even been credited with killing crickets and saving the crops around the Great Salt Lake area.

The **pelican** is one of the largest marine birds, easily recognized by its large beak. It can catch fish up to several pounds by diving from the air and crashing through the surface of the water to gobble up some unsuspecting fish. Pelicans look like a plane totally out of control as they crash into the sea, and it is always amazing to see them shake themselves off and fly away to dive again. They have a habit of flying very low just in front of a moving swell or wave and riding on the updraft of air that it creates. They can soar for long distances this way, just inches off the water, without moving their wings.

The **frigatebird** is unusual because it is a marine bird, living and feeding over the open sea, yet it cannot swim. The oil that most water birds have on their feathers to prevent absorption of water is missing in the frigatebird. It flies over the water and picks things off the surface, but it would drown if it landed on the water.

Wilson's petrel is a small bird that flies some 32,000 kilometers a year to nest in the Antarctic where it is summer in December, and spends August off the Gulf Coast. Petrels try to stay in a summer season all year long. The author has seen large colonies of them nesting on icebergs in the Ross Sea, in the Antarctic. Later the same year, he saw them in Florida. They have one

Figure 22-6   Cormorants are large birds commonly seen in many coastal areas.

Figure 22–7 A bottlenosed dolphin, a marine mammal, highly popular from the time of Greek mythology to today's TV programs.

of the longest migrations of the bird world, along with the distinction of being thought to be one of the most numerous birds in the world.

THE CLASS MAMMALIA

This class includes most of what humans consider the higher forms of life, perhaps because humans are members of this group. The marine mammals are well-known to everyone through television and other media. They include the order **Cetacea** (whales and porpoises), the other **Carnivora** (seal-type animals), and the order **Sirenia** (manatees, sea cows, and dugongs). These animals are all warm-blooded, with body temperatures well above the temperatures of the sea around them. This causes a heat loss problem that must be solved; several anatomical modifications seem to do the job of regulating body heat adequately. The shape of the mammals gives a large internal volume-to-surface area ratio. They are rounded in cross section, and the larger they get, the better heat conservation they have because of the increased internal volume-to-surface area ratio. It is worth noting that all marine mammals are generally larger than most land mammals. Other modifications are obvious upon inspection of a specimen: a fatty layer located under a thick skin for heat conservation and reduced blood circulation to the skin areas where heat would be lost to the cold water.

The ability of these mammals to dive very deep has always amazed scientists. We could not understand why they did not get the ''bends'' as humans do if they dive for an extended time and then come directly up to the surface. We also could not understand how they could hold their breath so long with no ill-effects to brain tissue. In recent years, most of these answers have been discovered through research. With the capture of these animals and their placement in facilities like ''Sea World Park'' and various ''marineland'' underwater zoos, we have learned much. These underwater recreational parks have given scientists a laboratory for marine studies never available to them before. The whales were found not to get the bends because when they dive, they exhale instead of inhaling as we do. Consequently, they cannot absorb any excess nitrogen; there is none to absorb because they have almost no air in their lungs.

The body has made many modifications to conserve oxygen and to tolerate higher amount of waste products, such as lactic acid from muscle activity and carbon dioxide. A few of the modifications are the slowing of the heartbeat; the reduced blood flow to all organs and sections of the body that are not vital to life; an extremely high hemoglobin count in the blood and myoglobin in the muscle, both of which retain high levels of oxygen; and very high tolerance to the by-products of metabolism, as mentioned above.

Order Cetacea: Whales, Porpoises, and Dolphins

These magnificent animals wander through the oceans of the earth over great ranges from the polar regions to the tropics. Most of them winter in warmer water, where they give birth to their young, and then in the spring return to colder polar waters, where the young are weaned. The spring polar waters are so rich in krill that they are perfect feeding grounds for these filter-feeding giants of the sea. The **blue whale** is the largest animal known to have lived on earth. The young are 8 meters long at birth, and a 150-ton adult will be nearly 30 meters long. By feeding on the planktonic forms, being grazers rather than predators, they conserve energy. The **sperm whale** is an example of a predatory form, feeding on squid and fish. It must move much faster and be more agile than the blue whale to catch its food. They are large animals, but only about one-half as long as a blue whale and one-third the weight. Smaller still are the **killer whales.** They grow to 8 or 9 meters and, until recent years, were thought to be ferocious killers. Now that humans have had the opportunity to catch them and work with them, we find that they are very intelligent and do not seem to pose any danger to people at all. With scuba diving as common as it is now, it was inevitable

Figure 22–8 A fur seal, a marine mammal suborder Pinnipedia, feeding exclusively on fish.

that face-to-face encounters would happen, and so they have. Whales have swum up, looked at the diver, and then gone about their business. The diver generally takes the encounter less calmly, but as yet no one has been hurt by a whale.

One basic difference in body form between cetaceans and fish is the tail. The mammal's tail is horizontal, and the swimming motion is basically up and down, combined with a bit of rotary motion. The tail of the fish is vertical, and the swimming motion is from side to side. A whale can swim at a speed of 10 knots for extended periods of time and can reach 20 knots for short periods. Fish are able to reach higher speeds, but do not often travel so far over a long period of time. Whales seem to move steadily along their migration route when they travel.

The whales and porpoises are divided into two suborders on the basis of whether they have teeth or not. The ones without teeth are called **baleen** whales (suborder Mysticeti) and are the filter feeders. In this group are the blue, humpback, and California gray whales, among others. Whalers named the gray whale and its close relations of the family Balaenidae, the "right whales." The name came from the fact that they did not sink when they were killed by harpoon. They were the "right whales" to get, as they were not lost by sinking when they were dead. The gray whale was protected by

Figure 22-9   This fin whale skull, as exhibited in a museum, is typical of the fine displays available to the public in many cities.

Figure 22-10   This sea lion shows the streamlined form common to the diving mammals. The stream of bubbles rolling up its back were exhaled by the animal to allow it to descend more easily.

law some years ago when their population was nearly depleted by whalers and has responded well to its protection by increasing the population back to a sizable number once again.

The whales with teeth are members of the suborder Odontoceti. This group has a single blowhole, or nostril, on top of the head where the filter feeders have a double nostril. This group also uses a type of sonar or echo location organ to navigate and find food in the water. In tanks these animals can swim through hoops and perform tricks even when their eyes are covered. The beaked whales are called **bottlenose** whales and get quite large, up to 32 meters in the northern Pacific species.

The bottlenose dolphin is the common form found in places like Sea World in San Diego, California and other underwater parks or zoos. **Porpoises** are found in all seas and even in rivers. They are known to move up the Amazon and Yanzi rivers for several hundred miles. The terms *porpoise* and *dolphin* are used by most people interchangeably, but the word *porpoise* generally refers to the short-beaked forms that have a rounded snout instead of the pronounced beak of the bottlenose types. The largest of the toothed whales is the sperm whale, made famous in the novel *Moby Dick*. It can dive deeper, as far as we know, than any other mammal, over one-half mile. The sperm whale, along with the blue whale, are two of our most endangered species. If we do not stop whaling, they will become extinct.

Order Carnivora:
Sea Otters, Seals, Sea Lions, and Walruses

Because of their shore-based life style, these animals are well-known to the public. They occur in one form or another on almost all shores of the world and in all zoos. They perform on television and in the movies as well as on stages in night clubs and circuses. The most common performer is the **California sea lion.** This animal has large flippers and can walk (waddle) across land better than most other marine mammals. These "eared" types include the fur seals of Pribilof Islands off Alaska, which are starting to breed on San Miguel Island off central California, as well as the southern fur seal found circumpolar on land masses that approach the Antarctic Circle (South America, Africa, New Zealand, etc.). The hair seals do not have the large forward flippers and are very clumsy on land. They also lack the external ear of the sea lions. The largest of this group is the elephant seal, which can be as long as 6 meters. Like the sea lions, the seals also have at least one species that lives in fresh water. It is the smallest of this group, only about 1 meter, and is found only in Lake Baikal, Siberia. The most common member of the seals, the **harbor seal,** is found in most parts of the world. It is small, friendly and quite often easily approached by humans. The crabeater seal will filter-feed much the same as the baleen whales and is more an open sea or open pack ice animal than many other forms. The Weddell seals of the Antarctic crawl out on the ice floes to have their young or find cracks in pressure ridges in the ice and come up from underneath the ice, miles from the open sea, to bear young where they are unmolested. In years of quick freezes, some are trapped out of water when the sea freezes over. Taylor Dry Valley in the Antarctic is famous for the seals found there, some of whom have been dead for 2,000 years. It is thought that they crawled up the valley when they were trapped by a quick freeze and could not return to the sea.

The **walrus** is in a separate group because it has flippers like the sea lions but no ears like the seals. Walrus also have tusks of ivory. The walrus is a bottom feeder and digs in the mud with its tusks for clams. The ivory tusks are considered valuable, and hunters have killed off the majority of the walrus population. They are rare but now protected and may make a comeback.

The **sea otters** are quite different in that they have legs with fingers in front and webbed hind feet for swimming. They are almost like a weasel. Nearly extinct because of hunting, they have come back strongly and are rather prolific now. They are found on the West Coast as far south as central California, where they have destroyed the abalone beds by eating them. They can be seen from shore, swimming on their back with an abalone or sea urchin resting on their stomach as they pick it apart with their front legs and eat it. They were hunted for their fine fur, which is prized for its beauty in coats and shawls.

Order Sirenia: Manatees, Sea Cows, and Dugongs

These animals are not often seen, and they live mainly in shallow areas, particularly where there is an influx of fresh water. They eat plants. The dugongs are nearly extinct or at least greatly reduced because they have been taken for their meat. They are Asiatic and African in range and do not occur in North or South America.

The **manatee** is found in the general area of Florida and the Caribbean as well as on the west coast of central Africa. The author has had the pleasure of swimming with these large mammals and found them quite calm and unafraid. The greatest population-reducing factor for this animal, like so many others, is that its environment is being destroyed as the human population increases.

In Sebastian Inlet on the east coast of Florida, the author watched several manatees lift their heads out of the water and feed on the overhanging mangroves. The heavy boat traffic in populated areas cause many injuries and deaths to the manatees as they are often cut by the propeller blades of

Figure 22–11   Research ships are often in rough seas. All gear on deck must be tied down and all on-board laboratories secured in such a way that work may be carried on even though the ship is rolling heavily from side to side.

passing boats. High-speed boats and skippers of low intelligence account for most of the accidents.

The sea cow was discovered in 1741 and totally killed off by 1768. When they became known, their habitat was several islands in the northern Pacific area. Killed for hides, food and oil, they no longer exist.

Order Carnivora

Some forms like the polar bears spend as much time at sea as many of the so-called marine forms. The polar bears swim and dive for food and are often found 15 kilometers from the nearest shore or ice floe; some have been reported as far as 150 kilometers offshore.

REVIEW QUESTIONS

1. List some of the adaptations the sea turtles have had to make in order to live in the marine environment.
2. How do the penguins differ from most other water birds?
3. How do marine mammals solve the heat-loss problem and maintain their body temperature?
4. The whales are divided into two major groups. Give the characteristics of each.
5. Why is the walrus placed in a separate group?

SUGGESTED READINGS

Ambrose, W. G. 1984. "Influence of Residents on the Development of a Marine Soft-Bottom Community." *Journal of Marine Research.* 42:633–654.

Amos, W. H. 1969. *Life in Bays.* National Audubon Society, Nature Program. Garden City, NY: Doubleday.

Barrington, E. J. W. 1979. *Invertebrate Structure and Function,* 2nd ed. New York: John Wiley & Sons.

Bates, M. 1960. *The Forest and the Sea.* New York: Signet Science Library Book.

Beebe, W. 1928. *Beneath Tropic Seas.* New York: Putnam's.

Berrill, N. J. 1966. *The Life of the Ocean.* New York: McGraw-Hill.

Brown, Jr., F. A.; Hastings, J. W.; and Palmer, J. D. 1970. *The Biological Clock: Two Views.* New York: Academic Press.

Carson, R. 1951. *The Sea Around Us.* Boston: Houghton Mifflin; New York: Oxford University Press, 1961.

Chaplin, C. G. 1972. *Fishwatchers' Guide to West Atlantic Coral Reefs.* Valley Forge, PA: Harrowood Books.

Coker, R. E. 1962. *This Great and Wide Sea.* New York: Harper & Row. (Torchbook).

Daugherty, A. E. 1972. *Marine Mammals.* Sacramento, CA: Department of Fish and Game.

Dawson, E. Y. 1944. *The Marine Algae of the Gulf of California.* Los Angeles: University of Southern California Press.

Detwyler, T. R. 1971. *Man's Impact on Environment.* New York: McGraw-Hill.

Edwards, C. E. 1971. *Persistent Pesticides in the Environment.* Cleveland, OH: Chemical Rubber Company Press.

Fell, B. 1975. *Introduction to Marine Biology.* New York: Harper & Row.

Fretter, V., and Graham, A. 1976. *A Functional Anatomy of Invertebrates.* New York: Academic Press.

Fry, D. H., Jr. 1973. *Anadromous Fish of California.* Sacramento, CA: Department of Fish and Game.

Furlong, M., and Pill, V. 1970. *Starfish.* Edmonds, WA: Ellison Industries.

Gilmartin, M. 1958. "Some Observations on the Lagoon Plankton of Eniwetok Atoll." *Pacific Science.* 12:313–316.

Gotshall, D. W. 1977. *Fishwatchers' Guide.* Monterey, CA: Sea Challengers.

Gotshall, D. W. 1982. *Marine Animals of Baja California.* Monterey, CA: Sea Challengers–Western Marine Enterprises.

Halstead, B. W. 1959. *Dangerous Marine Animals.* Cambridge, MD: Cornell Maritime Press.

Harrison, Peter. 1983. *Seabirds: An Identification Guide.* Boston: Houghton Mifflin.

Hay, J. 1965. *The Run.* Garden City, NY: Doubleday.

Herald, E. S. 1961. *Living Fishes of the World.* Garden City, NY: Doubleday.

Hill, M. M. 1963. *The Sea.* Vols. I and II. New York: Interscience Publishers, John Wiley & Sons.

Hood, D. W. (ed.) 1971. *Impingement of Man on the Oceans.* New York: Interscience Publishers, John Wiley & Sons.

Idyll, C. P. 1971. *Abyss: The Deep Sea and the Creatures That Live in It.* New York: Crowell.

Idyll, C. P. 1972. *Exploring the Ocean World: A History of Oceanography,* rev. ed. New York: Crowell.

Jackson, S. W. 1971. *Man and the Environment.* Dubuque, IA: Wm. C. Brown Company.

Johnson, M. E., and Snook, H. J. 1935. *Seashore Animals of the Pacific Coast.* New York: MacMillan.

Keen, A. M. 1971. *Sea Shells of Tropical West America.* Stanford, CA: Stanford University Press.

Kelley, D. W. 1966. *Ecological Studies of the Sacramento–San Joaquin Estuary.* California Fish & Game, Fisheries Bull. No. 133.

Kirpichnikov, V. S. 1981. *Genetic Bases of Fish Selection.* New York: Springer-Verlag.

Lanting, F., and Stegner, P. 1985. *Islands of the West.* San Francisco: Sierra Club Books.

Lewis, J. R. 1964. *The Ecology of Rocky Shores.* London: English Univeristy Press.

MacGinitie, G. E., and MacGinitie, N. 1949. *Natural History of Marine Animals.* New York: McGraw-Hill.

Marchant, J. and Prater, T. 1986. *Shorebirds: An Identification Guide.* Boston: Houghton Mifflin.

Meske, C. 1985. *Fish Aquaculture.* New York: Pergamon Press.

Newell, N. D. 1972. "The Evolution of Reefs." *Scientific American.* 228(6):54–69.

North, W. J. 1963. *Ecology of the Rocky Seashore Environment in Southern California and Possible Influences of Discharged Wastes.* Internal Conference Water Pollution 7(6/7):721–736.

Parrott, A. W. 1957. *Sea Anglers' Fishes of New Zealand.* London: Hodder and Stoughton.

Peterson, R. T. 1961. *A Field Guide to Western Birds.* Boston: Houghton Mifflin.

Peterson, R. T. 1980. *A Field Guide to the Birds East of the Rockies,* 4th ed. Boston: Houghton Mifflin.

Pirie, R. G. 1977. *Oceanography,* 2nd ed. New York: Oxford Press.

Potter, V. R. 1971. *Bioethics: Bridge to the Future.* Englewood Cliffs, NJ: Prentice-Hall.

Reise, Karsten. 1985. *Tidal Flat Ecology.* New York: Springer-Verlag.

Ricketts, E., and J. Calvin. 1968. *Between Pacific Tides,* 4th ed. revised by J. Hedgpeth. Stanford, CA: Stanford University Press.

Schultz, G. A. 1969. *The Marine Isopod Crustaceans,* Dubuque, IA: Wm. C. Brown.

Sea Frontiers—International Oceanographic Foundation. Miami, FL.

Siegfried, W. R. et al. 1983. *Antarctic Nutrient Cycles and Food Webs.* New York: Springer-Verlag.

Smith, H. W. 1961. *From Fish to Philosopher.* Garden City, NY: Doubleday.

Storer, T. I. 1972. *General Zoology.* New York: McGraw-Hill.

Sumich, J. L. 1976. *An Introduction to the Biology of Marine Life.* Dubuque, IA: Wm. C. Brown.

Tait, R. V. 1981. *Elements of Marine Ecology.* Boston: Butterworth.

Thiruvathukal, J. B., and McCormick, J. M. 1976. *Elements of Oceanography.* Philadelphia: W. B. Saunders.

Thorson, G. 1971. *Life in the Sea.* New York: McGraw-Hill.

Villee, C. A. et al. 1984. *General Zoology,* 6th ed. New York: Saunders College Publishing.

Vogt, F. 1982. *Fish Farming: Its Possible Future Structure.* Proceedings 2nd International Seminar on Energy Conservation and Renewable Energies in the Bio-Industries. New York: Pergamon Press.

Index